DARWINISM COMES TO AMERICA

DARWINISM COMES TO AMERICA

RONALD L. NUMBERS

HARVARD UNIVERSITY PRESS
Cambridge, Massachusetts, and London, England

Printed in the United States of America

Second printing, 1999

Library of Congress Cataloging-in-Publication Data

Numbers, Ronald L.
Darwinism comes to America / Ronald L. Numbers.
p. cm.
Includes bibliographical references and index.
ISBN 0-674-19311-3 (cloth : alk. paper)
ISBN 0-674-19312-1 (pbk. : alk. paper)
1. Evolution—Religious aspects—Christianity—History of doctrines. 2. Theology, Doctrinal—United States—History.
I. Title.
BT712.N85 1998
231.7'652'0973—dc21
98-16212

To

My friend Jon H. Roberts

Author of the best book on Darwinism in America

CONTENTS

INTRODUCTION: DARWINISM, CREATIONISM, AND INTELLIGENT DESIGN

The publication in 1859 of Charles Darwin's epoch-making book *On the Origin of Species by Means of Natural Selection* touched off a national debate that continues to divide American society. Scarcely a week passes without some evolution-related story appearing in the news: religious leaders declaring the scientific legitimacy of biological evolution, politicians expressing their belief in divine creation, local school boards wrangling over the teaching of origins, professors being ordered to refrain from questioning evolution in the classroom, state legislatures debating whether to fire teachers who present evolution as a fact, biology textbooks carrying disclaimers denying the factual basis of evolution, scientists claiming that they have discovered evidence of intelligent design in the natural world, and public opinion polls showing that nearly half of all Americans believe in the recent special creation of the first humans.

Early Reactions

At first Americans reacted cooly to Darwinism (a term commonly used as a synonym for organic evolution). As we will see in Chapter 1, American naturalists embraced biological evolution gingerly, and those who did accept it tended to downplay the importance of Darwin's preferred mechanism of natural selection operating on random variations. As long as the scientific community remained skeptical about the merits of Darwinism, theologians could remain on the sidelines, confident that speculations about monkeys becoming men would never be taken seriously as science. By the mid-1870s, however, most American naturalists who expressed themselves on the subject were speaking out positively, and by the close of the decade only a handful of prominent scientists continued to regard Darwinism as a false theory.

Darwin's success in convincing fellow naturalists of the truth of evolution prompted more and more religious leaders, such as James McCosh, the president of Princeton College, to take a public stand. Theological liberals in the Protestant camp fairly quickly adapted their reading of Scripture and their doctrinal beliefs to accommodate biological evolution, but most theological conservatives, representing the majority of Americans, viewed Darwinism, especially when applied to humans, as erroneous, if not downright dangerous. They feared that the notion of "might makes right" would undermine Christian morality and that tracing human genealogy back to apes would invalidate the concept of humans being created in the image of God. With few exceptions, however, even the most literalistic Bible believers accepted the antiquity of life on Earth as revealed in the paleontological record. They typically did so either by interpreting the days of Genesis 1 as vast geological ages (the day-age theory) or by inserting a series of catastrophes and re-creations or ruins and restorations into an imagined gap between the first two verses of the Bible (the gap theory). By the close of the nineteenth century virtually the only Christians writing in defense of the recent appearance of life on Earth and attributing the fossil record to the action of Noah's flood were Seventh-day Adventists, a Fundamentalist group numbering fewer than 100,000 members. Until the 1970s this uncompromisingly literal reading of Genesis, developed and popularized by the Adventist "geologist" George McCready Price, generally went by the name of flood geology.[1]

Like Protestants, Catholics split along the progressive-conservative divide, with most prelates and priests remaining on the latter side. In the 1890s Father John Zahm, a priest-scientist at the University of Notre Dame, took the lead in trying to harmonize Catholicism with a theistic version of Darwinism, but the Vatican effectively silenced him in 1897. The next year his book *Evolution and Dogma* appeared on the Index of forbidden books; this proscription cast a theological shadow over evolution among American Catholics for years to come. American Jews overwhelmingly rejected evolution till the mid-1870s, when some Reform rabbis began to urge its acceptance. By the early 1890s evolution had established itself in the Reform community, though traditional Jews often expressed skepticism.[2]

Evolution and Antievolution before Sputnik

Little organized opposition to evolution appeared before the early 1920s, when Fundamentalist Christians, led by the Presbyterian layman and three-time Democratic presidential candidate William Jennings Bryan, launched a state-by-state crusade to outlaw the teaching of human evolution in public schools. By the end of the decade they had succeeded in only three states—Tennessee, Mississippi, and Arkansas—but the ruckus they raised retarded the dissemination of evolutionary ideas in American classrooms for over three decades. The antievolution campaign of the 1920s attracted enthusiasts around the country, in the North as well as in the South, in urban as well as in rural areas, but it garnered little support among America's scientific elite. Its leading scientific authorities were Arthur I. Brown, an obscure Canadian surgeon whose handbills touted him as "one of the best informed scientists on the American continent"; S. James Bole, a science teacher at the Fundamentalist Wheaton College who had earned a master's degree in education with a thesis on penmanship in an Illinois school district; Harry Rimmer, a Presbyterian preacher and self-styled "research scientist" who had briefly attended a homeopathic medical school; and the self-trained George McCready Price, whom the journal *Science* identified as "the principal scientific authority of the Fundamentalists."[3]

Antievolutionists in the 1920s may have agreed on the evils of Darwinism, but they disagreed spiritedly over the correct interpretation of Genesis 1. As one frustrated creationist observed in the mid-1930s, Fundamentalists were "all mixed up between geological ages, Flood geology and ruin, believing all at once, endorsing all at once." As long as they remained split over the meaning of Genesis, he reasoned, they could scarcely expect to convert the world to their creationist way of thinking. The 1930s and 1940s witnessed multiple attempts to create a united Fundamentalist front against evolution, but each one failed because of the intransigence of the various partisans, especially the flood geologists, who refused to compromise on the recent appearance of life on Earth and the geological significance of Noah's flood. In 1941, at the invitation of the president of the Moody Bible Institute, a group of five evangelical Christian scientists met in Chicago to establish the American Scientific Affiliation (ASA), which initially opposed evolution but soon came to accept organic development over time, punctuated by divine interventions, especially for the creation of matter, life, and humans. More liberal Christians increasingly identified evolution as simply God's method of creation and ignored the problem of reconciling science and Scripture.

Meanwhile, during the same years, biologists, after decades of disagreeing over the mechanism of evolution to the point of fostering reports of Darwinism lying on its "death-bed," began to forge a common explanation of evolution, which came to be known, perhaps misleadingly, as the modern or neo-Darwinian synthesis. Geneticists, taxonomists, and paleontologists, who had long worked virtually isolated from one another, finally began interacting—and agreeing on the centrality of natural selection in the evolutionary process. In doing so, they repudiated other evolutionary explanations, particularly ones that gave evolution the appearance of having a purpose. This created, in the words of the historian-biologist William B. Provine, an "evolutionary constriction" that squeezed any talk of supernatural design out of biological discourse. "The evolutionary constriction," he asserts, "ended all rational hope of purpose in evolution," thus making belief in Darwinism the functional equivalent of atheism. Many evolutionists remained devout Christians and Jews, but it became increasingly difficult to do so on the basis of the scientific evidence for evolution.[4]

The Creationist Revival

The evolutionary constriction scarcely influenced the content of high school biology textbooks until after 1957, when the Soviet Union successfully launched Sputnik into space, greatly embarrassing the American scientific establishment. Politicians and science-policy experts quickly pinpointed the inferior scientific education of Americans as the underlying cause of the country's slide to second place in the space race. To remedy the situation, the federal government began pouring large amounts of money into improving science textbooks for high school students. In biology, where leading practitioners were complaining that "one hundred years without Darwinism are enough," the funds went to the Biological Sciences Curriculum Study (BSCS), which produced a series of texts featuring evolution as the centerpiece of modern biology. When these unabashedly proevolution texts descended on American classrooms in the early 1960s, they produced howls of protest from conservative Christians, who regarded the BSCS books as an ungodly "attempt to ram evolution down the throats of our children."[5]

Just as the BSCS controversy was breaking, two Fundamentalists, John C. Whitcomb, Jr., an Old Testament scholar, and Henry M. Morris, a hydraulic engineer, brought out a book called *The Genesis Flood*, which presented Price's flood geology as the only acceptable interpretation of the first chapters of Genesis. Their insistence on beginning with a literal reading of the Bible and then trying to fit science into that context, rather than constantly accommodating the Bible to the findings of science, struck a responsive chord with many concerned Christians. In substantial, though undetermined, numbers they abandoned the once-favored day-age and gap theories, which allowed for the antiquity of life on Earth, accepting instead the strict creationism of flood geology, which limited the history of life to no more than 10,000 years and affirmed creation in six twenty-four-hour days.

Two years after the appearance of *The Genesis Flood* Morris and nine other like-minded creationists banded together to form the Creation Research Society (CRS). The members of the CRS, all possessing some scientific or technical training at the graduate level, attacked not only evolution, as their intellectual forebears had done in the 1920s, but any compromise with theories of ancient life. Many members, including Morris himself, insisted that God had created the entire universe, not just earthly life, within the past 6,000 years or so. In contrast to the antievolutionists of the 1920s, who could claim no well-trained scientists of their own, the ten founders of the CRS included five biologists with earned doctorates from major universities, a biochemist and an engineer with Ph.D. degrees, two biologists with master's degrees, and a tenth member who pretended to have an M.A. in geology.

These young-Earth creationists, as they came to be called, proved highly effective in promoting Price's flood geology among conservative Christians. The San Diego–based Institute for Creation Research, which Morris had established in 1972, served as unofficial headquarters. About 1970, to help gain a platform for their views in public school classrooms—and to endow them with a measure of scientific respectability—the proponents of flood geology renamed Price's model of Earth history "scientific creationism" or "creation science." The sequence and timing of key events, such as a recent special creation and subsequent worldwide flood, remained the same, but all direct references to biblical characters and places, such as Adam and Eve, the Garden of Eden, and Noah and his ark, disappeared from the stripped-down narrative. Within a decade or two the tireless proselytizers for scientific creationism had virtually co-opted the generic creationist label for their hyperliteralist views, which only a half-century earlier had languished on the margins of American Fundamentalism. People who called themselves creationists during the last quarter of the twentieth century typically assumed that most listeners would identify them as believers in a young Earth.

The explanation for this dramatic shift in creationist thinking is difficult to nail down. Developments such as the evolution offensive launched by the BSCS help to explain the robust revival of creationism in the post-Sputnik period, but they scarcely account for the dramatic shift among Bible-believing Christians from old-Earth to young-Earth interpretations of Genesis 1. Facile generalizations about educational deprivation and cultural alienation simply will not suffice. Highly educated citizens may have been more likely than their less well-trained neighbors to subscribe to evolution, but a quarter of those Americans who professed belief in the recent special creation of the first humans had graduated from college. Within the evangelical community of Christians, Fundamentalists have displayed greater enthusiasm for scientific creationism than their Pentecostal brethren and sisters; yet few students of American religion would argue that the former are more socially alienated or economically depressed than the latter.

Whatever the reasons for the efflorescence of antievolutionism in the late twentieth century, the prodigious popularity of scientific creationism among conservative Christians almost certainly related more to theological than social impulses. Many converts were attracted by the creation scientists' insistence on giving the Bible priority over science. As believers who took the Bible as literally as possible, they found the young-Earth creationists' nonfigurative reading of the days of creation, the genealogies of the Old Testament, and the universal deluge of Noah to be especially appealing. No longer did they have to *assume* (as day-age advocates did) that Moses meant "ages" when he wrote "days" in Genesis 1; nor did they have to *assume* (as gap theorists did) that Moses, without explanation or comment, skipped over the longest period of Earth's history—between the creation "in the beginning" and the far later Edenic creation—simply to accommodate Scripture to science.

The theological factors that encouraged adoption of creation science varied among and within denominations. Independent Baptists, Missouri Lutherans, and Seventh-day Adventists, to name three of the religious groups most receptive to creation science, each possessed distinctive motivations for embracing it. Premillennial Baptists and Adventists (but not amillennial Lutherans), who interpreted the prophecies of Revelation, the last book of the Bible, as indicating an apocalyptic end of the world associated with the Second Coming of Christ, tended to see an intimate link between the beginning described in the first book of the Bible and end-time events. As the Baptist Henry Morris once observed, "If you take Genesis literally, you are more inclined to take Revelation literally." The Adventists (but not Baptists and Lutherans) possessed an extra-biblical endorsement of creation science in the divinely inspired writings of their prophetess, Ellen G. White. Though all three traditions read the Bible through literalistic lenses, the Missouri Lutherans (but not the Baptists or Adventists) well into the twentieth century went so far as to defend Ptolemaic astronomy, which placed Earth rather than the sun in the middle of the solar system.

Balanced Treatment

In the early 1980s state legislatures across the United States debated a creationist-inspired model bill that called for the balanced treatment of "evolution-science" and "creation-science" in public schools. Two states, Arkansas and Louisiana, enacted this proposed legislation into law, but the new statutes quickly encountered judicial opposition. In 1982, after a trial that brought more attention to the creation-evolution controversy than any event since Clarence Darrow confronted William Jennings Bryan in the Scopes trial of 1925, a federal judge in Little Rock declared the Arkansas law to be an unconstitutional breach of the wall separating church and state. Five years later the U.S. Supreme Court, after hearing a case from Louisiana, upheld a lower court decision that creation science served a religious, not scientific, purpose. However, one justice, writing for the majority, left the schoolhouse door open a crack for creationism to slip through. "Teaching a variety of scientific theories about the origins of humankind to schoolchildren," he wrote, "might be validly done with the clear secular intent of enhancing the effectiveness of science instruction."[6]

Since the 1920s the scientific establishment had paid little attention to the snipings of creationist critics, but the creationist successes (and near successes) of the early 1980s finally roused them from their apathy. Organizations from the National Academy of Sciences to little-known local science societies, fearing the loss of public patronage and cultural authority, denounced those who challenged their conclusions and sought to adulterate science with religion. Just how much influence their fulminations had on the voting of state lawmakers is hard to assess. In at least one state, creationist educators rather than evolutionary biologists brought about the defeat of the model bill. A scientist at the University of Oklahoma told of witnessing how a legislative committee in that state reached its decision to oppose the balanced-treatment act. At a public hearing in early 1981 a joke-cracking, down-home school superintendent from a rural district begged the legislators to "leave us alone—we know what we're doing. We're not teaching evolution—we're teaching biblical creation." The bill under consideration required that creation and evolution be given equal time if either were taught; because most Oklahoma schools were teaching only creation, the bill, seen as another "example of big government telling the local school boards what to do," would force them to expose students to evolution. Needless to say, the bill (and evolution) went down to defeat.[7]

Creationism in the 1990s

The Supreme Court's decision effectively ended efforts to mandate the inclusion of creationism in public school curricula, but it did little to slow down creationist initiatives to undermine evolution. Instead of agitating for balanced-treatment acts at the state level, creationists refocused much of their energy on individual schools and school districts, where in many instances considerable support for creationism already existed. In the early 1990s the National Center for Science Education (NCSE), which monitored creationist endeavors throughout the country, warned that people unfamiliar with precollegiate education would "be surprised at the amount of official antievolutionism that is found there, especially among administrators." In the fall of 1992 the Center drew attention to "a sharp surge upwards" in creationist attacks on evolution. These often took the form of calling for downgrading the status of evolution from "fact" to "theory" or for presenting students with "evidence against evolution," a notion the director of the Center, Eugenie C. Scott, dismissed as "merely 'scientific' creationism in sheep's clothing."[8]

Some educators employed novel solutions to solve the recurring evolution problem. In response to complaints about the inclusion of evolutionary cosmology in elementary school textbooks, the superintendent of schools in Marshall County, Kentucky, ordered that the offending two pages be glued together. The Cobb County school district in suburban

Atlanta, Georgia, went directly to the publisher of a troublesome fourth-grade text and asked that a chapter entitled "The Birth of Earth" be deleted. Modern electronic publishing allowed Macmillan/McGraw Hill, the publisher, to excise seventeen pages, thereby producing a custom-made text exclusively for the students of Cobb County.[9]

In Alabama the state school board in 1995 voted six to one in favor of inserting the disclaimer shown in the box on page 10 in all biology textbooks used by the state. Biology textbooks in Alabama subsequently began arriving from the publishers with that message pasted into the front. The Republican governor, Fob James, who presided over the board, strongly backed the disclaimer, saying that he personally believed the biblical account of the origin of life to be true.[10]

In the mid-1990s controversies over creationism erupted not only in Georgia, Kentucky, and Alabama but in Virginia, Pennsylvania, New Hampshire, Ohio, Indiana, Michigan, Wisconsin, New Mexico, California, and Washington. Tennessee legislators defeated a bill, at first expected to "blast through the House Education Committee like a rocket," that would have allowed the firing of any teacher who presented evolution as fact rather than theory. Such activity prompted one frustrated anticreationist to exclaim that "creationism is like a vampire, and every time you think the thing is finally dead, someone pulls the damned stake out again."[11]

As a Republican candidate for the presidency of the United States in 1980, Ronald Reagan had insisted that "if evolution is taught in public schools, creation also should be taught." In 1995 the Republican presidential candidate Pat Buchanan adamantly denied any kinship with simian ancestors: "I don't believe it is demonstrably true that we have descended from apes. I don't believe it. I do not believe all that." He *did* believe that parents had "a right to insist that Godless evolution not be taught to their children." During the 1990s various state Republican parties added creationist planks to their platforms. And in all regions of the country—North, South, East, and West—creationists stood for election to local school boards. They often won.[12]

Support for creationism ran deep in North American society. Despite the nearly unanimous endorsement of naturalistic evolution by leading biologists, a Gallup poll in 1993 showed that 47 percent of Americans continued to believe that "God created man pretty much in his present form at one time within the last 10,000 years," and an additional 35 percent thought that the process of evolution had been divinely guided.

A MESSAGE FROM THE ALABAMA STATE BOARD OF EDUCATION

This textbook discusses evolution, a controversial theory some scientists present as a scientific explanation for the origin of living things, such as plants, animals and humans.

No one was present when life first appeared on earth. Therefore, any statement about life's origins should be considered as theory, not fact.

The word "evolution" may refer to many types of change. Evolution describes changes that occur within a species. (White moths, for example, may "evolve" into gray moths.) This process is microevolution, which can be observed and described as fact. Evolution may also refer to the change of one living thing to another, such as reptiles into birds. This process, called macroevolution, has never been observed and should be considered a theory. Evolution also refers to the unproven belief that random, undirected forces produced a world of living things.

There are many unanswered questions about the origin of life which are not mentioned in your textbook, including:

- Why did the major groups of animals suddenly appear in the fossil record (known as the "Cambrian Explosion")?
- Why have no new major groups of living things appeared in the fossil record for a long time?
- Why do major groups of plants and animals have no transitional forms in the fossil record?
- How did you and all living things come to possess such a complete and complex set of "Instructions" for building a living body?

Study hard and keep an open mind. Someday, you may contribute to the theories of how living things appeared on earth.

Only 11 percent subscribed to purely naturalistic evolution. (Seven percent expressed no opinion.) Fifty-eight percent of the public favored teaching creationism in the schools. In Canada, which had experienced comparatively little controversy over origins, 53 percent of adults rejected evolution.[13]

In 1986, during a visit to New Zealand, the American paleontologist and anticreationist Stephen Jay Gould assured his hosts that scientific creationism was so "peculiarly American" that it stood little chance of "catching on overseas." His colleague Richard C. Lewontin seemed to agree. "Creationism is an American institution," he declared, "and it is not only American but specifically southern and southwestern." So it may have seemed at the time, but scientific creationism was already traveling far beyond the borders of the United States, enjoying growing popularity in Europe, Asia, and the South Pacific. In 1980 Australian antievolutionists established the Creation Science Foundation (CSF) in Queensland, where for a period in the 1980s creation appeared in the state syllabus for secondary schools. Within a short time the CSF became the world's second leading center for the propagation of scientific creationism (after Morris's Institute for Creation Research). In the mid-1990s its star speaker, Kenneth A. Ham, opened an international creationist organization in Florence, Kentucky (near Cincinnati), as "an outreach of the CSF." The Korea Association of Creation Research, also founded in 1980, expanded so rapidly that it, too, established branches in the United States.[14]

Even in Islamic cultures such as Turkey's creationism made extensive inroads. In the 1980s the ministry of education translated several creation-science books into Turkish and distributed them to teachers throughout the country. In a report on evolution sent to Turkish educators, the minister of education dismissed Darwinism as a handmaiden of materialism based on "nothing but some interpretations and guesswork." He recommended that biology textbooks "provide all of the evidence in favor of and against the theory of evolution," so that Turkish youth could "gain the habit of objective and scientific thinking."[15]

Varieties of Evolutionism and Creationism

Although popular accounts of the creation-evolution controversies of the 1990s often characterized them as a bipolar debate between naturalistic evolutionists and supernaturalistic creationists, opinions on origins actually spanned a wide intellectual spectrum. On one end were the *naturalis-*

tic evolutionists, often atheists or agnostics (such as the Oxford biologist Richard Dawkins and the Tufts philosopher Daniel C. Dennett), who saw no evidence of, or need for, a Creator God. Nearer the center were the *theistic evolutionists,* often devout Christians (such as the physicist Howard J. Van Till, of Calvin College, and many members of the evangelical American Scientific Affiliation), who saw little or no evidence of God in nature but who, for theological reasons, believed that God had created the world by means of evolution. On the other side of the center were the *intelligent-design theorists* (such as the Australian biochemist-physician Michael Denton, the Berkeley law professor Phillip E. Johnson, the Lehigh biochemist Michael J. Behe, and the editors of the journal *Origins & Design*), who rejected naturalistic evolution and claimed to see evidence of an Intelligent Designer in the complexity of nature, but who often accepted the antiquity of life on Earth. At the opposite end of the spectrum from the naturalistic evolutionists, beyond a mixed group of *old-Earth creationists* (such as the California-based astronomer Hugh Ross), were the *scientific creationists,* typically Fundamentalist Christians (such as Henry M. Morris and most members of the Creation Research Society), who compressed the entire history of the universe into little more than 6,000 years and postulated a divine creation in six literal days.

The "point man" for naturalistic evolution in the 1990s was Dawkins, author of such books as *The Blind Watchmaker* (1986), described on the dust jacket as perhaps "the most important book on evolution since Darwin." The title, a reference to the role of natural selection in creating organized complexity, left little doubt of Dawkins's position:

> Natural selection, the blind, unconscious, automatic process which Darwin discovered, and which we now know is the explanation for the existence and apparently purposeful form of all life, has no purpose in mind. It has no mind and no mind's eye. It does not plan for the future. It has no vision, no foresight, no sight at all. If it can be said to play the role of watchmaker in nature, it is the blind watchmaker.

In an oft-quoted statement, Dawkins praised Darwin for making "it possible to be an intellectually fulfilled atheist," and he repeatedly went out of his way to bait creationists, all of whom he believed to be "ignorant, stupid or insane." He dismissed the first chapters of Genesis as just another creation myth "that happened to have been adopted by one particular tribe of Middle Eastern herders" and theistic evolution as a superfluous attempt to "smuggle God in by the back door." No wonder one of Dawkins's patrons, Charles Simonyi, a rich Microsoft executive who endowed a special professorship for Dawkins at Oxford, fondly called his beneficiary "Darwin's Rottweiler," a reference to the nineteenth-century evolutionist and agnostic Thomas H. Huxley, famous as "Darwin's bulldog."[16]

If Dawkins played the role of point man for late-twentieth-century naturalistic evolutionists, Daniel C. Dennett gladly served as their hatchet man. In a book called *Darwin's Dangerous Idea* (1995), which Dawkins warmly endorsed, Dennett portrayed Darwinism as "a universal solvent, capable of cutting right to the heart of everything in sight"—and particularly effective in dissolving religious beliefs. The most ardent creationist could not have said it with more conviction, but Dennett's agreement with them ended there. He despised creationists, arguing that "there are no forces on this planet more dangerous to us all than the fanaticisms of fundamentalism." Displaying a degree of intolerance more characteristic of a fanatic Fundamentalist than an academic philosopher, he called for "caging" those who would deliberately misinform children about the natural world, just as one would cage a threatening wild animal. "The message is clear," he wrote: "those who will not accommodate, who will not temper, who insist on keeping only the purest and wildest strain of their heritage alive, we will be obliged, reluctantly, to cage or disarm, and we will do our best to disable the memes [traditions] they fight for." With the bravado of a man unmindful that only 11 percent of the public shared his enthusiasm for naturalistic evolution, he warned parents that if they insisted on teaching their children "falsehoods—that the Earth is flat, that 'Man' is not a product of evolution by natural selection—then you must expect, at the very least, that those of us who have freedom of speech will feel free to describe your teachings as the spreading of falsehoods, and will attempt to demonstrate this to your children at our earliest opportunity." Those who resisted conversion to Dennett's scientific fundamentalism would be subject to "quarantine."[17]

Evolutionary ideologues such as Dawkins and Dennett made headlines for their hard-nosed views, but not all naturalistic evolutionists took such a draconian line. For example, Stephen Jay Gould, a self-described agnostic, expressed dismay at such tough rhetoric and welcomed any signs of rapprochement between naturalistic evolutionists and theists. He celebrated Pope John Paul II's recognition in 1996 that the theory of evolution is "more than a hypothesis," even though the pontiff rejected materialistic evolution and insisted on the special creation of the soul. Gould knew that the Dawkinses and Dennetts of the world would say, "C'mon, be honest; you know that religion is addlepated, superstitious, old-fashioned b.s.; you're only making those welcoming noises because religion is so powerful, and we [that is, evolutionists] need to be diplomatic in order to assure public support and funding for science," but he refused to countenance their intolerance. Scientists who claimed that Darwinism had disproved the existence of God should have "their knuckles rapped for it"—as should those who claimed that evolution was God's method of creation. In the late 1980s Gould even trained a young-Earth creationist and doctoral candidate, Kurt P. Wise, in his laboratory and, though bemused by his beliefs, always treated him with respect, even after he joined the faculty of the Fundamentalist William Jennings Bryan College in Dayton, Tennessee.[18]

Theistic evolutionists, representing 40 percent of the American population, occupied a spectral range from borderline naturalistic evolutionism to progressive creationism, but the most prominent Protestant advocates in the 1990s leaned more toward the former than the latter. Ever since its creation in the early 1940s, the American Scientific Affiliation (ASA) has served as the primary forum for evangelical Christian scientists interested in issues related to creation and evolution. During its early years the ASA leaned toward creationism, but that began to change in the late 1950s. During the Darwin centennial of 1959, at a time when even the most progressive evangelical scientists were still limiting the scope of evolution, Walter R. Hearn, a biochemist in the ASA, pushed for uncompromising evolution, which he regarded as God's method of creation. Though a fervent theist, he denied that scientists could discover direct evidence of God's activity by looking through a microscope. By the 1990s views such as his had come to dominate ASA thinking.[19]

Unfortunately for the ASA and theistic evolutionists generally, the din of debate between naturalistic evolutionists and scientific creationists often drowned out their irenic voices. The evangelical physicist Van Till, one of the most vocal theistic evolutionists of the 1980s and 1990s, described evolution as "an ordinary natural process—a process that is not fundamentally different in character or status from other natural processes, such as a summer sunrise, a winter snowstorm, the blooming of a flower, or the birth of a child." Although evolution did "not require the introduction of phenomena that go beyond the ordinary pattern of material behavior," it nevertheless occurred within "the Creator's domain of action" and thus was not "inherently naturalistic." This brand of theistic evolution appealed to many Protestants (as well as Catholics and Jews) who sought middle ground in the battle between creation and evolution, but it elicited only scorn from naturalistic evolutionists and scientific creationists, who agreed on few matters, noted one observer, except that "theistic evolution is woefully—even perniciously—confused." One side disliked it because it was theistic; the other, because it was evolution. But mostly it was ignored. After all, noted the antievolutionist Phillip Johnson condescendingly, theistic evolutionists occupied "just a few backwater positions" in the debate over origins.[20]

Scientific (or young-Earth) creationism pretty much dominated creationist discourse after the 1960s, but large pockets of evangelical Christianity remained loyal to (if comparatively silent about) the once-dominant old-Earth models associated with the day-age and gap interpretations of Genesis 1. According to Henry M. Morris in 1995, "the most influential current scientist" then writing in defense of the day-age interpretation of Genesis 1 was Hugh Ross, an astronomer who believed that Morris's scheme was not only bad science but bad exegesis. Perhaps the most visible defender of the gap theory in late-twentieth-century America was the lusty charismatic TV evangelist Jimmy Lee Swaggart, whose Pentecostal tradition had long detected two different creation events in the first chapter of Genesis.[21]

Intelligent Design

Just as scientific creationists had captured the fancy of journalists in the 1980s, intelligent-design (ID) theorists grabbed headlines in the 1990s with their bold calls for rewriting the basic rules of science and their claims of having found indisputable evidence of God. The intellectual roots of the ID movement go back centuries, but its contemporary incarnation dates from the mid-1980s. In 1984 three Protestant scientists, Charles B. Thaxton, Walter L. Bradley, and Roger L. Olsen, brought out *The Mystery of Life's Origin*, in which they attributed the complex process of originating life to a divine Creator. The most striking feature of their book was not its text but its foreword, contributed by Dean H. Kenyon, a Catholic professor of biology at San Francisco State University and the coauthor of a major text on the chemical origins of life. Confessing that he no longer believed in naturalistic evolution, Kenyon joined the authors of the book in identifying "a fundamental flaw" in current theories about the origins of life. "A major conclusion to be drawn from this work," he wrote, "is that the undirected flow of energy through a primordial atmosphere and ocean is at present a woefully inadequate explanation for the incredible complexity associated with even simple living systems, and is probably wrong."[22]

Two years later Michael Denton, an expatriate Englishman living in Australia, wrote an iconoclastic book, *Evolution: Theory in Crisis* (1986), questioning the validity of neo-Darwinism and arguing that evidence of divine design exists in nature. Although he had grown up in a religiously conservative family, he no longer maintained ties with organized religion or harbored any sympathy for young-Earth creationism. He did, however, see humans and other organisms as the products of God-ordained laws of nature.[23] Neither *The Mystery of Life's Origin* nor *Evolution: Theory in Crisis* attracted much attention among mainstream scientists, but they both helped to lay the intellectual foundation for the ID movement of the 1990s.

The first ID book to reach a wide audience—and one of the first explicitly to adopt the intelligent-design slogan—appeared in 1989 under the title *Of Pandas and People: The Central Question of Biological Origins.* The authors, Dean H. Kenyon and Percival Davis, designed this slim, illustrated volume as a supplement to high school biology texts written from a Darwinian point of view. Using six case studies, they compared Darwinian and ID explanations to see which better matched the scientific data. Not surprisingly, intelligent design—defined as a frame of reference that "locates the origin of new organisms in an immaterial cause: in a blueprint, a plan, a pattern, devised by an intelligent agent"—always won. An appended "Note to Teachers" went out of its way to distinguish ID theory from religious Fundamentalism and scientific creationism, but skeptical evolutionists regarded the theory as simply a "creationist alias." By 1996 at least two states, Alabama and Idaho, and a number of local school districts had evaluated the suitability of the book for adoption in public schools. Most, if not all, had recommended against its use.[24]

In 1991 the infant ID movement received a big boost from an unlikely source: a University of California, Berkeley, law professor, Phillip E. Johnson. A few years earlier the Presbyterian Johnson had stumbled across Dawkins's *The Blind Watchmaker* and discovered, as he put it, that the argument for evolution was more rhetorical than factual. Being a lawyer, he recognized the practice all too well. In a book titled *Darwin on Trial* he sought to expose the soft underbelly of Darwinism by critically examining the evidence for the blind-watchmaker thesis. Both in that book and in a subsequent work, *Reason in the Balance: The Case against Naturalism in Science, Law and Education* (1995), Johnson disclosed what he saw as the core problem with naturalistic evolution: its unwarranted assumption that naturalism is the only legitimate way of doing science. This bias, he argued, unfairly limited the range of possible explanations and ruled out, a priori, any consideration of theistic factors.[25]

Denton hailed *Darwin on Trial* as "the best critique of Darwinism I have ever read," and the governor of Alabama used his discretionary funds to send copies of the book to all biology teachers in the state. Gould, in contrast, dismissed it as "scarcely more than an acrid little puff," unworthy of a serious response. In a scathing review in *Scientific American,* Gould insisted that "science can work only with naturalistic explanations; it can neither affirm nor deny other types of actors (like God) in other spheres (the moral realm, for example)." When the editor of *Scientific American* denied Johnson's request for "equivalent space" to respond to Gould, the ID camp saw the editor's action as a confirmation of their suspicions of official discrimination against theistic views. When, about the same time, Kenyon's department at San Francisco State University ordered him to quit teaching "creationism" (he had introduced students to the evidence against Darwinism and for intelligent design), the ID camp acquired a martyr.[26]

Until the mid-1990s no major academic or trade press had published a work supporting intelligent design or, indeed, creationism of any kind. That changed in 1996, when the Free Press of New York released Michael J. Behe's *Darwin's Black Box: The Biochemical Challenge to Evolution.* Behe, a Catholic biochemist at Lehigh University, had first become aware of the alleged difficulties of Darwinism through reading Denton's book. Later he read Johnson's *Darwin on Trial,* which confirmed his growing doubts about the adequacy of naturalistic evolution to explain molecular life. When a reviewer of Johnson's book in the journal *Science* treated it harshly, Behe rushed to the lawyer's defense with a letter to the editor. He subsequently began exchanging correspondence with Johnson himself and drafting his own book-length reply to naturalistic evolutionists such as Dawkins, whom he regarded as "the best modern popularizer of Darwinism around." In *Darwin's Black Box* Behe argued that biochemistry had "pushed Darwin's theory to the limit . . . by opening the ultimate black box, the cell, thereby making possible our understanding of how life works." The "astonishing complexity of subcellular organic structure" led him to conclude—on the basis of scientific data, he asserted, "not from sacred books or sectarian beliefs"—that intelligent design had been at work. "The result is so unambiguous and so significant that it must be ranked as one of the greatest achievements in the history of science," he declared without a trace of false modesty. "The discovery [of intelligent design] rivals those of Newton and Einstein, Lavoisier and Schroedinger, Pasteur and Darwin."[27]

As newspapers and magazines spread the news of Behe's discovery, he won recognition as a modern-day William Paley, the most famous natural theologian of the early nineteenth century. The influential evangelical magazine *Christianity Today* honored *Darwin's Black Box* with its "Book of the Year" award for 1997. Like so many other ID theorists, Behe distanced himself as far as possible from the scientifically disreputable young-Earth creationists, going so far as to concede the possibility that the universe had been around for billions of years and that life on Earth had developed from a common ancestor. But such disclaimers scarcely deterred critics from deriding his views as "thinly veiled creationism." The great nemesis of theistic science, Dawkins, chided Behe on television for lazily relying on intelligent design when he should have gone looking for scientifically acceptable explanations of his data.[28]

By the mid-1990s the ID theorists, many of whom had been collaborating since the 1980s, were coalescing into an institutionalized movement: organizing conferences, establishing a center, and publishing a journal, *Origins & Design*. The journal's masthead listed the leading lights of ID theory, with Denton, Thaxton, Kenyon, Johnson, and Behe all serving on the editorial advisory board. A trio of young Christian philosophers of science—Paul A. Nelson, Stephen C. Meyer, and William A. Dembski—headed the editorial office. Many of these same men were connected with the generously endowed Center for the Renewal of Science and Culture, affiliated with the Discovery Institute in Seattle. Although individually they espoused a wide range of views on origins (from Denton's and Behe's virtual theistic evolutionism to Nelson's young-Earth creationism), collectively they staked out a position between theistic evolutionism—"American evangelicalism's ill-conceived accommodation to Darwinism"—and scientific creationism, American Fundamentalism's ill-conceived effort to base science on Scripture. Their goal was "an intellectual revolution" that would redraft the basic rules of science to include nonnaturalistic explanations of phenomena. Unlike some mid-century creationists who felt the need to form a united front on creation before challenging evolution, the ID theorists set aside the construction of a specific creation model while they tried to mount a unified attack against Darwinism. "When the Goliath [of naturalistic evolution] has been tumbled," they reasoned, "there will be time to work out more details of how creation really did occur."[29]

"In so pluralistic a society as ours," Dembski once asked rhetorically, "why don't alternative views about life's origin and development have a legitimate place in academic discourse?" He already knew the answer: the scientific establishment's alleged bias toward atheistic materialism, a bias acknowledged by at least a few of America's scientific elite. The distinguished evolutionary biologist Richard Lewontin, for example, worried in the 1990s that the public might actually believe what Dawkins and other careless popularizers told them about evolution, which often rested on "unsubstantiated assertions or counter-factual claims." In an unvarnished statement that conformed precisely to what the ID theorists had been claiming, he described the workings of the modern scientific mind: "We take the side of science *in spite* of the patent absurdity of some of its constructs, *in spite* of its failure to fulfill many of its extravagant promises of health and life, *in spite* of the tolerance of the scientific community for unsubstantiated just-so stories, because we have a prior commitment, a commitment to materialism."[30]

As they no doubt anticipated, the intelligent-design crowd took a beating from all sides: scientific creationists, theistic evolutionists, and, of course, naturalistic evolutionists. Although a few young-Earth creationists, such as Nelson and Wise, applauded the effort to discover evidence of God in nature, the leaders of creation science, despite believing in divine design, never warmed up to ID theory. One scientist on the staff of the Institute for Creation Research faulted the advocates of ID for their "lack of reliance on the literal statements of Scripture and the construction of alternative models of origin, which involve long periods of years." Henry M. Morris, the grand old man of scientific creationism, admired the efforts of ID theorists to refute Darwinism but lamented their apparent lack of concern for theological niceties. In embracing the paleontological record of life and death on Earth, as many did, they seemed "indifferent to the fact that this means accepting a billion years of a suffering, dying biosphere before Adam's fall brought sin and death into the world."

Morris predicted that, despite having compromised on the plain meaning of the Bible, the proselytizers for ID theory would find no more favor with naturalistic evolutionists than he himself had.[31]

The theistic evolutionists in the ASA, who also believed in a divinely designed world, remained skeptical of ID theory for other reasons. In an editorial on intelligent design in the ASA journal, *Perspectives on Science and Christian Faith,* the Gordon College chemist J. W. Haas, Jr., surmised that "most evangelical observers—especially working scientists—are deeply skeptical." Though supportive of theistic world views, they balked at being "asked to add 'divine agency' to their list of scientific working tools." To rely on intelligent design to explain complex biological organisms was, said Haas quoting Dawkins, "a pathetic cop-out of [one's] responsibilities as a scientist." Besides, Haas noted, ID theorists rarely applied their methods to disciplines outside of biology, leaving "the rest of us as physicists, chemists, mathematicians, or geologists . . . to go our 'godless' ways in spite of the complexities we face at the quantum level or with the weather."[32]

Reactions to ID theory among naturalistic evolutionists were overwhelmingly negative. One annoyed critic no doubt captured the feelings of many when he described it as "the same old creationist bullshit dressed up in new clothes." When the Jewish magazine *Commentary* in 1996 published a version of ID theory by the mathematician and novelist David Berlinski, letters of protest poured onto the editor's desk. Dennett ridiculed Berlinski's stylish essay as "another hilarious demonstration that you can publish bull—t at will—just so long as you say what an editorial board wants to hear in a style it favors." Another reader characterized Berlinski's "intuitions about the Design of the World as neither more nor less reliable than those of flat-earthers, goat-entrail readers, or believers in the Oedipus complex."[33]

Historians are notoriously unreliable prophets, but it seems safe to predict that America will continue to witness spirited, indeed acrimonious, debates over the scientific, theological, and political consequences of evolution for the foreseeable future. As much as some people—scientists no less than Fundamentalists—might like to dictate what their fellow citizens should believe in regard to origins, it appears unlikely that they will succeed in a constitutional, democratic, and divided republic, where only about one in ten adults subscribes to naturalistic evolution and a majority believe the book of Genesis to be inerrant. As long as the Bible remains the most trusted and widely read text in America and scientists maintain their considerable cultural authority, consensus, even if desirable, seems unlikely.[34]

More than six decades have passed since Bert James Loewenberg wrote his pioneering Harvard dissertation, "The Impact of the Doctrine of Evolution on American Thought."[35] During that time, and especially since the Darwin centennial in 1959, a number of superb studies of Darwinism in America have appeared. Among the best are three biographies: of Asa Gray by A. Hunter Dupree, of Louis Agassiz by Edward Lurie, and of Joseph LeConte by Lester D. Stephens.[36] Edward J. Pfeifer, Peter J. Bowler, and Mary P. Winsor have explored the responses of various American scientists to Darwinism, while Jon H. Roberts, James R. Moore, and David N. Livingstone have done the same for people concerned about the religious implications of Darwinism.[37] Christopher P. Toumey, George E. Webb, Edward J. Larson, and I have all explored antievolutionism in America.[38] In the meantime, historians of science and religion have rejected or refined the once-dominant cliché about science and religion being in a state of perpetual war.[39] Given this wealth of literature—and I have not mentioned many good books, articles, and unpublished dissertations—what more is there to say? Plenty, I think.

In the chapters that follow I begin by investigating the diverse uses of such basic terms as Darwinism, evolutionism, antievolutionism, and creationism in the American context. Rather than defining these terms according to present-day understandings—such as insisting that Darwinism implied an emphasis on natural selection and that creationism demanded that Earth be no older than about 6,000 years—I look at how such terms were used in the past and how their meanings have changed over time.[40] As I show in Chapters 1 and 2, most Americans seem to have employed Darwinism and evolutionism interchangeably, and until well into the twentieth century they rarely used the term creationism. Over the years these terms proved to be so malleable and covered such a wide range of beliefs that it is often difficult to distinguish an evolutionist from an antievolutionist.

To get a fairly accurate sense of scientific responses to Darwinism, I have eschewed generalizations based on the views of a handful of select persons who expressed themselves volubly on the subject in favor of a prosopographical analysis of the views of the eighty American naturalists elected to the National Academy of Sciences between 1863 and 1900. In this way I have been able to identify more antievolutionists than previously known and more naturalists who either took no interest in evolution or studiously avoided the topic. I have also found evidence to suggest that the so-called neo-Lamarckian school remained relatively small and that neo-Darwinism, as the term was used by contemporaries, never gained a foothold in the United States. Although few American naturalists identified their reasons for accepting organic evolution, I am convinced by the clues they did leave that the perceived bankruptcy of special creation as a scientific explanation played at least as great a role as the positive evidence for evolution.

A decade ago the historian Peter Bowler brought out an iconoclastic little book exposing the ballyhooed Darwinian revolution as "a historical myth." He argued that although evolution won the struggle against special creation, Darwin's theory of natural selection lost to other proposed mechanisms in the competition to explain evolution.[41] My research (discussed in Chapter 1) leaves me convinced of the validity of Bowler's claims about the marginality of natural selection in the nineteenth century, but I remain willing to use the phrase Darwinian revolution because of Darwin's catalytic role, in America at least, in discrediting what he called "the dogma of separate creations."

Historians of Darwinism in America have paid far too little attention, I think, to regional and denominational variations. As I show in Chapter 3, the common misrepresentation of two or three celebrated incidents in the American South has led to the portrayal of that region as distinctively hostile to Darwinism. Antievolution sentiment may indeed have been stronger there than elsewhere in the country (though we do not know for sure), but southerners were far more tolerant of evolution there than one would suspect from reading most historical accounts. Most southern states, for example, rejected antievolution laws in the 1920s, and those that forbade the teaching of Darwinism limited the ban to human evolution. In Chapter 4, on the trial of John Thomas Scopes for violating the Tennessee antievolution law, we can see the distorting effects not only of regional prejudice but of willful ignorance of the beliefs of Fundamentalists.

Historical studies of religious reactions to Darwinism have typically focused on the mainstream Protestant denominations to the virtual exclusion of the so-called sects. This concentration has led to the conclusion (largely valid for mainstream Protestantism) that one denomination's response to Darwinism was much like another's.[42] Outside the mainstream, distinctive beliefs made a great deal of difference. Because of the visions of their founding prophetess and their observance of the seventh-day Sabbath as a memorial of the Creation, Seventh-day Adventists (discussed in Chapter 5) diverged from other theologically conservative Christians in insisting on the recent creation of life and a flood at the time of Noah that deposited the fossil-bearing rocks. In the 1970s this once distinctively Adventist interpretation of Genesis became popular in Fundamentalist circles as "scientific creationism" or "creation science." In contrast to the hyperliteralistic Fundamentalists, the experientially oriented Holiness and Pentecostalism traditions (discussed in Chapter 6) remained less vulnerable to the arguments for young-Earth creationism. They also grew explosively in the twentieth century. By the 1990s Pentecostals and their charismatic brothers and sisters had become the new mainstream of Christianity, with about one fourth of the two billion Christians in the world professing that faith.

I hope in the following chapters not only to illuminate some of the dark corners of the history of evolution in America but to dispel a number of myths and misperceptions that still cling to narratives of Darwinism in the United States. Among them are:

- that Darwin's scientific influence in America resulted largely from his success in convincing biologists of the truth of natural selection;
- that American scientists who accepted evolution divided into opposing camps of neo-Lamarckians and neo-Darwinians;
- that Darwinism created a national religious crisis in the late nineteenth century;
- that antipathy to evolution has been characteristic of the American South;
- that Fundamentalists generally disliked and feared science;
- that Fundamentalists and conservative evangelicals could not accommodate the accumulating evidence from geology and paleontology about the great age of Earth because the Bible taught otherwise;
- that antievolutionists typically condemned all evolution, not just the evolution of humans;
- that William Jennings Bryan betrayed his Fundamentalist friends and supporters at the Scopes trial by admitting to the antiquity of life on Earth;
- that social class played a greater role than theology in influencing religious responses to evolution.

As we will see, each of these generalizations is dubious if not downright wrong.

1

DARWINISM AND THE DOGMA OF SEPARATE CREATIONS: THE RESPONSES OF AMERICAN NATURALISTS TO EVOLUTION

Speculations about organic evolution had been circulating among literate Americans throughout the first half of the nineteenth century, but before the appearance of Darwin's *Origin of Species* in 1859, few American naturalists besides Samuel S. Haldeman of Philadelphia had expressed any public sympathy for the so-called transmutation or development hypothesis, and even he stopped short of endorsing it. A few of his American colleagues shared his open-mindedness, but they tended to keep their heterodox views private. The publication of the *Origin* moved the topic of organic evolution out into the open and made it a subject of intense discussion. Scientific societies debated it; college students titillated one another with it; critics railed against it. By the early 1870s the paleontologist Edward Drinker Cope was declaring evolution to be an "ascertained fact," and within a few more years scientific opposition in North America had diminished to a whisper. Naturalists continued to argue about the adequacy of natural selection to account for evolution, but with few exceptions they, like Darwin, had forever turned their backs on the special creation of species.[1]

Nothing in the paragraph above is particularly novel or controversial. Those statements simply reflect the historical consensus that has developed since the publication of Bert James Loewenberg's pioneering 1933 article, "The Reaction of American Scientists to Darwinism," the first treatment of its kind by a trained historian. Unfortunately, Loewenberg, convinced that the "formidable triumvirate" of Louis Agassiz, Asa Gray, and James Dwight Dana dominated the post-1859 American scientific community, focused almost exclusively on the views of these three men to determine how American scientists responded to Darwin.[2] Subsequent studies, most notably by Edward J. Pfeifer and Peter J. Bowler, have greatly expanded the evidentiary base on which to make generalizations.[3] However, we still lack a systematic, broad-based survey of changing scientific opinions in America. We do not have, for example, satisfactory answers to such basic questions as which American naturalists converted to organic evolution, when they did so, why they did so, what evolutionary mechanisms they embraced, and what psychological and theological consequences resulted from their encounters with evolution.

Naturalists in the National Academy of Sciences

To help answer such queries, I have collected information about the experiences of the eighty American naturalists—biologists, geologists, and anthropologists—elected to the National Academy of Sciences (NAS) between its founding in 1863 and the end of the century (see the Appendix). The community of active naturalists in the latter part of the nineteenth century comprised far more than these eighty men, but this relatively large group serves as a practical base from which to examine scientific responses to Darwinism in the United States. It includes the great majority of the leading naturalists of the time, not just those who wrote at length about evolution. Although it reflects the prejudices of contemporaries, it avoids biases based on posthumous reputation. These men were respected by their peers, not necessarily by their successors.

A profile of this group shows that they ranged in age from the Yale patriarch Benjamin Silliman, born in 1779, to the anthropologist Franz Boas, born in 1858. Their births clustered in the second quarter of the nineteenth century:

1770s	1
1780s	0
1790s	4
1800s	8
1810s	12
1820s	14
1830s	18
1840s	15
1850s	8

The primary disciplinary interests of the Academy naturalists (and many of them worked in more than one field) were:

Anatomy, Physiology, and Medicine	8
Botany	9
Ethnology and Anthropology	3
Geology and Paleontology	34
Zoology	26

Geographically, they concentrated heavily in New England and the Middle Atlantic region, especially in the Boston and Washington areas, which, respectively, sent nineteen and eighteen members to the Academy. Calculating regional representation on the basis of time of residence after election to the Academy, we obtain the following distribution:

New England	30
Middle Atlantic	38
South	1
Midwest	7
West	2
Non-U.S.	2

The two nonresidents, the physiologist Charles-Edouard Brown-Séquard and the geologist William M. Gabb, spent little time in the United States after becoming academicians, but they were long-standing members of the American scientific community. Throughout the late nineteenth century the South remained strikingly underrepresented, with only one member, the internationally renowned herpetologist John E. Holbrook, of Charleston. Created by Congress during the Civil War, the Academy until 1872 imposed a loyalty oath on its members, which discouraged southern participation.[4] Besides, in the aftermath of the war, three current or future Academy members departed the South: Henry James Clark moved from Kentucky to Massachusetts, and Joseph LeConte left South Carolina for California, where he was joined a few years later by the former Mississippi geologist Eugene W. Hilgard. These two ex-southerners made up the entire western contingent in the Academy. Institutionally, the largest numbers came from Harvard University (thirteen), the U.S. Geological Survey (ten), Yale University (seven), and the University of Pennsylvania (five).

Darwinists and Anti-Darwinists

The first problem encountered in exploring how these American naturalists responded to Darwinism is definitional: What did it mean to be a

Darwinist or, for that matter, an anti-Darwinist? As Darwin scholars such as James Moore, David Hull, and Ernst Mayr have noted, these are not easy questions to answer, in part because of the complex and changing views of Charles Darwin himself.[5] In writing the *Origin,* Darwin simultaneously pursued two goals: overthrowing "the dogma of separate creations" and establishing natural selection as the primary, though far from exclusive, mechanism of change. Regarding the relative importance of these twin goals, he left no doubt. "Personally, of course, I care much about Natural Selection," he confided to an American correspondent; "but that seems to me utterly unimportant, compared to the question of Creation *or* Modification."[6] Despite occasional calls for more precise usage, many Americans indiscriminately employed the term Darwinism to refer to either (or both) of these goals, and they tended to use the words "evolution" and "Darwinism" interchangeably.[7] Thus the use of these terms tells little about the intended meaning.

Labeling opinions is difficult even for such well-known figures as the Harvard botanist Asa Gray, perhaps Darwin's preeminent American disciple during the first decades after the appearance of the *Origin.* Although Gray labored mightily to ensure that Darwin received "fair-play" in North America, and once described himself as "one who is scientifically, and in his own fashion, a Darwinian," he broke with Darwin on several key issues. He not only urged a "special origination" in connection with the appearance of humans and expressed skepticism about the ability of natural selection "to account for the formation of organs, the making of eyes, &c.," but proposed that the inexplicable organic variations on which natural selection acted be attributed to divine providence, a suggestion Darwin found scientifically unacceptable. No wonder Gray confessed to a friend that his theistic version of evolution was *"very anti-Darwin"*—or that the Princeton theologian Charles Hodge refused to credit Gray as a Darwinist.[8] Similarly, the Swiss-born paleobotanist Leo Lesquereux, a devout Lutheran who in 1864 announced his conversion to evolution in a personal letter to Darwin, continued into old age to regard "the first appearance" of organisms as lying outside the purview of science.[9]

Confusion also bedevils attempts to pin down what it meant to be an antievolutionist or what is now called a creationist (a label seldom used in the nineteenth century, as we will see in Chapter 2). In the *Origin* Darwin took aim at "the ordinary view of creation," but his exact target remained fuzzy. Was he referring to the biblically inspired Linnaean notion of the simultaneous creation of single pairs in one locale, which then multiplied and migrated to their eventual homes; or Charles Lyell's alternative suggestion of various "*centres* or *foci* of creation," separated spatially and temporally; or Louis Agassiz's proposal of repeated plenary creations, which stipulated that "species did not originate in single pairs, but were created in large numbers," in the habitats the Creator intended them to populate? Agassiz, whose earlier studies of glaciers had led him to postulate the existence of an ice age, believed that the geological record revealed a series of catastrophes and special creations by which the earth had been repeatedly depopulated and repopulated. For him, species were linked by the mind of the Creator, not by common descent from Eden or Ararat. When Asa Gray referred to "the commonly received doctrine" of creation, he explicitly identified it with this nonbiblical doctrine of Agassiz's.[10]

Just as Gray fails to conform to the stereotypical Darwinist, Agassiz scarcely fits the mold of a Bible-thumping creationist. If this suave Swiss immigrant attended church at all, it was with liberal Unitarians. Regarded in some quarters as an "infidel," he scoffed at the supposition that fossils represented "the wrecks of the Mosaic deluge" and dismissed the story of Adam and Eve as an "absurdity." During his long tenure on the Harvard faculty, he forbade any "tenet teaching" in the classroom.[11] He personally trained 20 percent of the Academy naturalists, virtually all of whom went on to accept organic evolution. Rather than seeing their mentor as a barrier to the spread of evolution in America, some of his former students portrayed him as a Darwinian John the Baptist, who "prepared the way for the theory of evolution" by inculcating the need for empirical investigation, by correlating the embryological development of living organisms with "the order of succession of their extinct representatives in past geological times," and by emphasizing "progressive development" in the organic world, for which even Darwin expressed appreciation. Shortly before his death, Agassiz confided to a former student that he would "have been a great fellow for evolution if it had not been for the breaks in the paleontological record."[12]

After Agassiz's death in 1873 the most prominent American antievolutionist in the scientific community—according to one hyperbolic report, the *only* "working naturalist of repute in the United States . . . that is not an evolutionist"—was the Princeton geographer-geologist Arnold Guyot. A charter member of the Academy, this theologically orthodox naturalist operated within a loosely biblical framework but experienced little difficulty accommodating the antiquity of Earth, the progression of the fossil record, and a local deluge. As a leading popularizer of the nebular hypothesis of the origin of the solar system, he assigned most of the work of creation to divinely ordained laws of nature. For years he resisted accepting the transmutation of one species into another, but in his final statement on the subject, published in 1884, he demanded only the special creation of matter, life, and humans. "Evolution from one of these orders into the other—from matter into life, from animal life into the spiritual life of man—is impossible," he declared, adding that "the question of evolution within each of these great systems—of matter into various forms of matter, of life into the various forms of life, and of mankind into all its varieties—remains still open." His close friend James Dwight Dana, who reached similar conclusions, suspected that Guyot had come "to accept, though with some reservation, the doctrine of evolution through natural causes."[13]

Creationism among Academy naturalists thus spanned a conceptual spectrum ranging from a virtual infinitude of miraculous interventions (Agassiz) to perhaps only three (Guyot). None of the Academy's critics of evolution defended a literal six-day creation or tried to squeeze Earth's history into a mere 6,000 years, and some of them, such as Guyot, could have passed as evolutionists. Conversely, several of the Academy's purported evolutionists clung to the divine origin of humans. As we have already seen, even Darwin's champion Asa Gray for years regarded the "special origination" of the first humans as a "great likelihood." Dana, after years of hesitation, came to accept "the derivation of man from an inferior species," but still insisted "that there was a divine creative act at the origin of man . . . as truly a creation as if it had been from earth or inorganic matter to man." The anthropologists Lewis Henry Morgan and Frederick Ward Putnam seem never to have accepted the Darwinian history of human development.[14] Within the National Academy of Sciences, as in American society generally, the boundary separating creationists from evolutionists long remained blurry.

The lack of a clear line of demarcation makes it virtually impossible to carry out the seemingly simple task of identifying the evolutionists in the Academy. Complicating the job is the refusal or apparent reluctance of many members to express themselves on what contemporaries and historians alike judge to be the major issue of the day confronting naturalists. The Catholic geologist James Hall, for example, refused to state his position, despite the prodding of friends and associates. Nineteen others apparently took so little interest in the subject that their biographers have seen no reason to mention it. Thus I can say nothing about the opinions on evolution of fully one quarter of the academicians; and an additional nine, who almost certainly accepted evolution, left few clues with which to reconstruct their beliefs. Of the sixty members for whom I possess some evidence, fifty-one seem to have accepted evolution, five or six (Louis Agassiz, George Engelmann, A. A. Gould, Arnold Guyot, Edward Hitchcock, and possibly John S. Newberry) apparently remained creationists of one kind or another, two (Agassiz's son Alexander and Thomas Sterry Hunt) adopted a skeptical, equivocal attitude, while one (the preacher-turned-geologist J. P. Lesley) flirted for a time with evolution but later denounced it as "the prevalent epidemic scientific superstition of the day."[15]

Because only a few of the Academy naturalists recorded the time of their conversion to evolution, it is extremely difficult to pinpoint when the community of Academy naturalists embraced Darwinism or claimed to do so. Three of the eighty—Samuel Haldeman, William Barton Rogers, and Asa Gray—seem to have accepted the transmutation of species before reading the *Origin*, but I have found no evidence that any of the others did so. And except for a couple of the youngest members of the group (Charles E. Beecher and Franz Boas), there is no evidence of anyone's adopting evolution after 1880. Of the thirty others for whom we can estimate the date of conversion, about half converted in the 1860s and the other half in the 1870s. Age clearly played an important role in influencing who joined the Darwinian camp and who did not. Of the thirteen Academy naturalists born before 1810, only three became evolutionists, and one of them (Lesquereux) did so with reservations. In contrast, the sixteen youngest members of the group all openly embraced evolution.[16]

Debating Darwinism

Although some American men of science had been evaluating the merits of evolution since the 1840s, Charles Darwin did not begin to influence the discussion until about 1857, when he confidentially shared with Asa Gray his distinctive ideas on transmutation. Copies of the *Origin* began reaching the East Coast during the last days of 1859, and before the year was out America's leading paleontologist, Joseph Leidy of Philadelphia, was promising to send Darwin some New World evidence "in support of the doctrine of selection." Leidy, who credited Darwin with bringing him "out of the darkness" like "a meteor [that] flashed upon the skies," also successfully maneuvered, in March of 1860, to elect the English naturalist to membership in the Academy of Natural Sciences of Philadelphia, an honor Darwin greatly appreciated.[17]

Meanwhile, in New England, Asa Gray was taking the lead in defending Darwin against Agassiz in the American Academy of Arts and Sciences while William Barton Rogers was assuming that responsibility in the Boston Society of Natural History. At a meeting of the latter society on February 15, 1860, Louis Agassiz dismissed Darwin's work as "an ingenious but fanciful theory," readily falsified by the persistence of certain organic forms over extensive periods of time. Rogers conceded that the persistence of types presented a "formidable objection" to Darwin's view, but he speculated that Darwin would probably defend himself by saying that "the vital characters of some animals fit them for resisting change and extinction better than more plastic natures." He then pressed Agassiz to name any vertebrate that "had ever been found in strata lower than the upper Silurian," whereupon Agassiz noted that "the highest representatives" of some animals are found in the lowest rocks. The exchange between Agassiz and Rogers continued for four meetings, at times drifting to such topics as the role of subsidence in the deposition of the Paleozoic strata of New York. The minutes of the meetings report no appeals to God or *ad hominem* remarks about simian ancestry, and Rogers described his "friendly contests" with Agassiz as "very courteous and good-natured." The local press ignored the exchange, and Rogers fretted that he was having "to do battle almost unaided."[18]

Rogers may have overdramatized his role as Darwin's lone defender, but not by much. For years most of the Academy naturalists, along with their colleagues generally, remained publicly silent on the topic of evolution. Despite the scientific excitement generated by the *Origin*, some of America's leading lights, such as the aging Benjamin Silliman, apparently never bothered to read the book, while others, such as Gray's botanical mentor, John Torrey, put off the task for several years. A *New York Times* reporter covering the 1860 annual meeting of the American Association for the Advancement of Science (AAAS) chided participants for giving "the propositions of Mr. Darwin a wide berth." In view of the obvious importance of his theory, such behavior looked "very much like either cowardice or incompetence." When the geologist John S. Newberry finally broached the topic, in his presidential address to the AAAS in 1867, he described Darwinism as "shaking the moral and intellectual world as by an earthquake." Nevertheless, he cautiously urged his colleagues to restrict Darwin's hypothesis to "its proper bearing upon the limits of variation, and the mooted question of 'what is a species?' " He comfortingly—but erroneously—assured his listeners of Darwin's "own acceptance of the doctrine of Revelation."[19]

Prodded at first by Agassiz's encouragement to study the Darwin question—in order to help show, as one student reported, "what is right and what is wrong in our pure and beloved science"—zoology students at Harvard soon began debating Darwinism. About 1862 Alpheus Hyatt, an invertebrate paleontologist, became the first of Agassiz's students to embrace the new theory. In recollecting the activities of the Harvard Natural History Society in the mid-1860s, when he was an undergraduate member, the botanist William G. Farlow described discussions as being "to a great extent about the origin of species, and, no matter what was the subject of the papers announced for the evening meeting, it was not often that we adjourned without dropping into a discussion of evolution. Few had really read Darwin's book, but all felt able to discuss the great scientific question of the day." Except for Agassiz's occasional barbs, talk of Darwinism generally remained outside of Harvard classrooms until the mid-1860s, and it did not enter the curricula of many other American colleges until the 1870s.[20]

Although the available evidence suggests that eighteen of the eighty Academy naturalists accepted evolution by the end of the 1860s, only a few publicly revealed their change of views. Gray and Rogers were the first to do so, followed in 1862 by J. D. Whitney, director of the California state geological survey, who tutored the legislators of California on the essentials of Darwinism. Hyatt liked to cite an 1866 paper of his on the parallels between fossil mollusks and present-day embryological development as an early, if frequently overlooked, contribution to evolutionary theory. But as Mary P. Winsor has pointed out, Hyatt said nothing in that paper to which Agassiz himself would have taken exception. Two years later, however, the twenty-eight-year-old vertebrate paleontologist Cope published an essay entitled "On the Origin of Genera," in which he explicitly tied his views to Darwin's work on the origin of species. Apparently the first of Agassiz's students to announce his conversion was Edward S. Morse, who openly professed his new faith before the Essex Institute in 1873. Just before his death that same year Agassiz ruefully conceded that the idea of organic development had won "universal acceptance" within the scientific community. The disappointed antievolutionist may have exaggerated the extent of the Darwinian victory, but he was not mistaken in believing that the tide of scientific opinion had turned decisively against the creationist views he so fervently held.[21]

Evolutionary Theories

But for what had so many of Agassiz's colleagues abandoned special creation? The obvious answer is evolution, but that says little about the great diversity of beliefs embraced by American evolutionists. As we have already seen, becoming an evolutionist in the eyes of contemporaries did not necessarily mean rejecting all acts of special creation. And it certainly did not mean accepting all that Darwin taught. Although Darwin in the *Origin* stressed the role of natural selection operating on minute accidental variations to produce new species slowly over time, he also appealed—from the first edition on—to a host of other factors. To compensate for his acknowledged "ignorance of the laws of variation," he increasingly invoked such mechanisms as the direct action of the physical environment and the inheritance of characters acquired through use and disuse—factors typically associated with the late French evolutionist Jean Baptiste Lamarck.[22]

From the beginning the Academy naturalists who adopted evolution tinkered with Darwin's scheme to suit their own intellectual or psychological needs. This was true even of those members who came the closest to accepting Darwin's own reconstruction of life on Earth. Gray, for example, suggested that God had directed the course of variations "along certain beneficial lines," while Rogers faulted Darwin for "ignoring entirely violent and sudden physical changes" and for downplaying "the gradual modification of species through external conditions." Yale's O. C. Marsh, who credited Darwin with inaugurating a new epoch in the history of science by speaking "the magic word—'*Natural Selection*,'" and who was regarded by some colleagues as a partisan of natural selection, nevertheless supplemented that mechanism with an inherent tendency toward brain growth.[23]

None of the Academy's evolutionists denied that natural selection played at least a "secondary" role in weeding out unfit organisms, but a number of them failed to see how natural selection could account for the *origin* of the fittest. Led by Edward D. Cope and Alpheus Hyatt, who joined the Academy in the early to mid-1870s, this group turned for a solution to the Lamarckian factors that Darwin had occasionally invoked but had not stressed. By 1876 the entomologist A. S. Packard, Jr., was referring to "an original and distinctively American school of evolutionists," whom a decade later he dubbed the neo-Lamarckians. Although the neo-Lamarckians held a number of views in common, and Cope, Hyatt, and Packard served at times as editors of the *American Naturalist*, the most recent historian of the so-called school, Theodore J. Greenfield, found "no evidence in their correspondence that they coordinated their efforts, assigned special tasks among themselves, or pursued their special areas in relation to other's choices."[24]

Best known for assuming the heritability of acquired characters, the neo-Lamarckians devoted far more of their energy to searching for natural, external causes of the origin of variations. "The influence of climate on variation has been studied to especial advantage in North America," Packard maintained, "owing to its great extent, and to the fact that its territory ranges from the polar to the tropical regions, and from the Atlantic to the Pacific." Opinions varied greatly among the neo-Lamarckians, but Cope and Hyatt tended to emphasize the role of "acceleration and retardation" in evolution. In one of the most lucid explications of these terms, Cope explained that

> there are two kinds of evolution, progressive and retrogressive; or, to use expressions more free from objection, by addition of parts, and by subtraction of parts. It is further evident that that animal which adds something to its structure which its parents did not possess, has grown more than they; while that which does not attain to all the characteristics of its ancestors has grown less than they. To express the change in the growth-history which constitutes the beginning of evolution, I have employed the terms "acceleration and retardation." . . . The origin of the fittest is then a result of either acceleration or retardation.

No wonder Edward Morse insisted on treating this theory as simply "a very plain case of natural selection."[25]

In one sense Cope and Hyatt were their own worst enemies. Unlike Darwin, who seduced readers with his luminous prose, both Americans repelled prospective followers by indulging their habit of mystifying their readers with arcane neologisms. Cope, for example, first combined "bathmism (growth-force), kinetogenesis (direct effect of use and disuse and environmental influences), and archaesthetism (influence of primitive consciousness)" into a unified theory before he "lost all contact with reality," as an erstwhile disciple put it, and began rapidly inventing even more polysyllabic labels: "Catagenesis, Bathmogenesis, Ergogenesis, Emphytogenesis, Statogenesis, and Mnemogenesis." Hyatt suffered from the same affliction. In writing Hyatt's biographical memoir for the National Academy of Sciences, an exasperated William Keith Brooks, who as a young zoologist had worked with Hyatt at the Boston Society of Natural History, confessed his inability to fathom the meaning of Hyatt's ruminations on evolution. "I have studied his more recent writings upon the subject with all the diligence that my great respect and admiration for him demanded, I have listened attentively when he has discussed his views in public, and I have had many private talks with him about them," he wrote plaintively, "but I do not understand them." Darwin, too, professed never to have comprehended Hyatt and Cope's theory of acceleration and retardation. "I have endeavoured, and given up in despair, the attempt to grasp their meaning," he told an American correspondent.[26]

Because most American naturalists who expressed themselves on the subject allowed for the action of at least some Lamarckian factors, one sometimes gets the impression (from primary and secondary sources alike) that the neo-Lamarckian "school" comprised the majority of American naturalists. But by that standard, Darwin himself was a neo-Lamarckian. If we count only those naturalists identified by their contemporaries or themselves as neo-Lamarckians, the group shrinks considerably. Within the Academy we find only seven: Cope, Hyatt, Packard, Charles Emerson Beecher, William H. Dall, Henry Fairfield Osborn, and Theodore N. Gill. All but Packard, an entomologist, and Gill, an ichthyologist, were paleontologists. By the late 1890s most of them (including Cope, Hyatt, Packard, Gill, and Osborn) had quit defending the inheritance of acquired characters as a fact.[27]

Until the 1880s the notion of inheriting acquired characters had seemed commonsensical, even to Darwin, who cited the well-known experiments of Brown-Séquard in its favor. This peripatetic physiologist, who joined the Academy in 1868 after several extended stays in the United States, had shown in the late 1850s that artificially produced epilepsy in guinea pigs could become hereditary in their offspring. Such powerful experimental evidence seemed to settle the question—until the German zoologist August Weismann mounted a sustained logical and experimental attack on the idea in the 1880s. Insisting that Lamarckian inheritance had never been demonstrated, he sought to show the vacuity of the theory by cutting off the tails of generations of mice and observing whether or not their offspring grew shorter tails. They didn't (just as 4,000 years of circumcision had not altered the appearance of Jewish penises). More important, Weismann insisted that the germ plasm, or hereditary material, of an organism could not be affected by somatic or bodily influences. This seemingly freed "Darwinism from all taint of Lamarckism" and made "selection the all-sufficient and, indeed, sole factor in species-forming." In 1888 the British biologist George J. Romanes called this reliance on the *Allmacht* of natural selection neo-Darwinism.[28]

The ensuing debate between Weismann's followers and his critics led contemporaries and, later, historians to characterize late-century evolutionists as being rent into hostile camps of neo-Darwinians and neo-Lamarckians. Rarely, however, do we find any American neo-Darwinians mentioned by name. For example, Lester Frank Ward in an address entitled "Neo-Darwinism and Neo-Lamarckism" in the early 1890s claimed that Weismann's "vigorous onslaught" on the inheritance of acquired characters "has probably aroused a greater amount of interest among scientific men than any other event that has transpired since the appearance of Darwin's *Origin of Species*"—but not a single American name appears among his neo-Darwinians. Similarly, Peter Bowler has asserted that neo-Darwinism "rapidly polarized the scientific community into two mutually hostile camps"—but he, too, assigns no Americans to the neo-Darwinian party.[29]

One of the rare turn-of-the-century identifications of American neo-Darwinians appears in David Starr Jordan and Vernon Lyman Kellogg's *Evolution and Animal Life* (1907), which names as prominent neo-Darwinians both Asa Gray and William Keith Brooks, a former student of Agassiz's who taught zoology at Johns Hopkins. Clearly, the inclusion of a theistic evolutionist such as the long-deceased Gray should be dismissed as a mistake. But Brooks was at least plausible. His biographer, Keith Benson, has described him as "the leading American neo-Darwinian in the late nineteenth century," and Brooks may have come as close as any of the Academy naturalists to sharing Weismann's views. However, he pointedly denied being influenced by Weismann, and he stopped far short of relying solely on natural selection to explain organic evolution. In the 1880s he was seeking to fashion a theory that lay "midway between Darwin's view of the origin of variation (as purely fortuitous) and the Lamarckian view (as resulting from the direct influence of the environment)." As late as 1896 he was describing natural selection as "no more than *a great but not the exclusive means* of adaptive modification." That is not the language of a Weismannian neo-Darwinian.[30]

Brooks's student E. B. Wilson, a cell biologist who spent much of his career at Columbia, is a second possible American neo-Darwinist. He repeatedly expressed admiration for both Weismann and the principle of natural selection. However, as Garland E. Allen has pointed out, Wilson faulted Darwin for placing "too much emphasis on evolution by the accumulation of small variations" and too much stress on "chance" in the origin of species. How, he asked late in life, can natural selection "explain the initial evolution of complicated organs, such as the eye, the ear or the heart which, it would seem, must have attained a certain degree of complexity before they could function at all"? He apparently never got over his aversion to the notion "that higgledy-piggledy can provide an adequate explanation of organic adaptations."[31]

Several Americans studied with Weismann at the University of Freiburg in the late nineteenth century, but, according to Frederick B. Churchill, none seems to have returned home to champion his mentor's neo-Darwinist project. Neo-Darwinism apparently flourished most in Great Britain, where it was commonly associated with the names of Alfred Russel Wallace and E. Ray Lankester. Although widely discussed in the United States, it never attracted much of a following. Among Academy members Weismann may have helped to undermine belief in the inheritance of acquired characters, but he seemingly won no converts to his strong claims for natural selection.[32]

Some Academy naturalists, such as Yale's William Henry Brewer and Harvard's Charles S. Minot, found Weismann's conclusions regarding the inheritance of acquired characters scientifically unpersuasive, but other critics focused more on the frightening social and moral implications of the German professor's teachings. "If Weismann and Wallace be right," warned Joseph LeConte, "then alas for all our hopes of race improvement—physical, mental, and moral!—for natural selection will never be applied by man to himself as it is by Nature to organisms. His spiritual nature forbids. Reason may freely use the Lamarckian factors of environment and of use and disuse, but is debarred the unscrupulous use of natural selection *as its only method.*" Henry Fairfield Osborn also worried about the consequences of Weismann's theory for human conduct. "If the Weismann idea triumphs, it will be in a sense a triumph of fatalism," he declared, because "each new generation must start *de novo*, receiving no increment of the moral and intellectual advance made during the lifetime of its predecessors."[33]

Such statements, linking human morality to evolutionary biology, appeared infrequently in the writings of the Academy naturalists, but they did crop up from time to time. One of the most vocal advocates of applying the teachings of Darwin to human society was Edward S. Morse, who cheered the progress of science in replacing religion as the determinant of social policy. "Man," he argued in his presidential address to the American Association for the Advancement of Science,

> now becomes an object of rigid scientific scrutiny from the new position which has shed such a flood of light upon the animals below him. His habits, behavior, the physical influences of his environment and their effects upon him, transmission of peculiarities through the laws of heredity,—all these factors are directly implicated in the burning questions and problems which agitate him to-day. Questions of labor, temperance, prison reform, distribution of charities, religious agitations are questions immediately concerning the mammal man and are now to be seriously studied from the solid standpoint of observation and experiment and not from the emotional and often incongruous attitude of the church.

The geologist-anthropologist John Wesley Powell, who shared Morse's rejection of traditional religion, nevertheless condemned efforts to apply "the methods of biotic evolution" to humans. "Should the philosophy of [Herbert] Spencer, which confounds man with the brute and denies the efficacy of human endeavor, become the philosophy of the twentieth century," he warned, "it would cover civilization with a pall and culture would again stagnate."[34]

If there were no neo-Darwinians in the Academy and only seven contemporaneously identified neo-Lamarckians, what did the rest of the forty-four probable evolutionists believe? Sixteen of them said so little on the subject that it is difficult to determine what they thought; and many others were so eclectic that it is hazardous to categorize their views. Nevertheless, some lumping might be useful if we keep in mind the deceptive nature of the assigned quantities and recognize that the categories are not mutually exclusive.

Eight or nine of the Academy naturalists tended to appeal to such factors as the effects of the environment and of use and disuse but were not specifically identified by contemporaries with the neo-Lamarckian school.[35] Six of the academicians, mostly geologists, stressed the role of environmental catastrophes or paroxysms in organic evolution. In 1877 the geologist Clarence King, addressing an audience at Yale's Sheffield Scientific

School, threw an "ignited bomb-shell" into the evolutionary camp by insisting that evidence from the American West, which he had personally surveyed, proved that each new modification of an organism "succeeded a catastrophe." In the same year that King delivered his iconoclastic speech, the naturalist Joseph LeConte, of the University of California, published a paper advocating what he called paroxysmal evolution. In what he regarded as "one of the most important" of his essays, he, too, argued for correlating "rapid changes of physical conditions and correspondingly rapid movement in evolution." This view no doubt attracted some support because, in the words of the clergyman-geologist George Frederick Wright, it did "not differ much in its phenomena from the old-style theory of special creation." Not surprisingly, the catastrophists tended to become the most enthusiastic admirers of the Dutch botanist Hugo de Vries, whose theory of evolution by mutation attracted considerable attention as an alternative to natural selection shortly after the turn of the century.[36]

Although neo-Lamarckians often invoked environmental changes to explain the production of new species, the philosopher Charles Sanders Peirce contrasted catastrophism not only with Darwinism but with neo-Lamarckism as well. "If evolution has been brought about in this way," wrote Peirce, "not only have its single steps not been insensible, as both Darwinians and Lamarckians suppose, but they are furthermore neither haphazard on the one hand, nor yet determined by an inward striving on the other, but on the contrary are effects of the changed environment." Thus, for him, catastrophists such as King and LeConte were not to be confused with neo-Lamarckians.[37]

So-called theories of orthogenesis, which depicted evolution occurring in an orderly direction, propelled by internal rather than external forces, provided another option for American evolutionists. Lamarck himself had postulated the existence in organisms of an inherent tendency toward greater complexity, and some of his American disciples followed his lead. In their hands orthogenesis typically took on a teleological coloration. However, shortly after the turn of the century the biologist Charles O. Whitman, who concurrently headed the department of zoology at the University of Chicago and directed the Marine Biology Laboratory at Woods Hole, Massachusetts, set out to rescue orthogenesis from its religious past and to reconcile it with natural selection. Having "found indubitable evidence of species-forming variation advancing in a definite direction," he urged his colleagues not to confuse the orderliness of nature with divine purpose. "In our aversion to the old teleology, so effectually banished from science by Darwin, we should not forget that the world is full of order, the organic no less than the inorganic," he advised. "Indeed, what is the whole development of an organism if not strictly and marvelously orderly?" For decades selectionists had tried in vain to account for the early stages of organs such as the eye. Orthogenesis now offered an explanation in terms of nonrandom variations. E. B. Wilson found Whitman's views somewhat attractive, but apparently no naturalist elected to the Academy in the nineteenth century concurred with Whitman in recognizing the orthogenetic process as "the primary and fundamental one" in evolution.[38]

Evolution and Religion

Theistic evolutionists, who saw the hand of God gently guiding the course of organic development, could be found in virtually all of the above groups, but eight Academy naturalists cannot be designated more precisely. Either they viewed evolution, in the words of the entomologist Samuel H. Scudder, primarily as "a law which is working out the plans of a Supreme Intelligence," or they left little evidence of their beliefs besides a statement of faith identifying evolution as God's method of creation. Asa Gray may have been the most prominent of this group, but none of his fellow theists in the Academy seems to have embraced his specific recommendation to attribute the direction of variation to divine guidance. As the editor of the *Independent* reported, Gray stood "nearly alone" in holding this ground. Peter Bowler has argued that theistic evolution was eclipsed in the late nineteenth century "by the rise of a naturalistic approach to the question of the origin of species." In the sense that references to the divine in the scientific literature became less visible, I would agree with him; but I suspect that at the time theistic evolution was undergoing privatization more than elimination. As early as 1860 Elliott Coues's father reported that he had been advised to delete an allusion to the Bible from a scientific manuscript, *"because a reference should never be made in scientific works to the Bible."*[39]

Bowler has also claimed that Darwinism "established a complete break between science and religion."[40] But little evidence from the lives of the Academy naturalists supports such a sweeping generalization. These men represented a broad spectrum of religious views, ranging from Baptist to atheist:

Catholic	3
Episcopal	4
Congregational	5
Presbyterian	6
Baptist	1
Lutheran	1
Methodist	1
Quaker	1
Theosophical	1
Mind Cure	1
Unitarian	6
Christian (denomination unknown)	4
Theist	6
Agnostic/Atheist	13
Unknown	27

I have found no evidence in either biographical or autobiographical accounts to suggest that a single one of these men severed his religious ties as a direct result of his encounter with Darwinism. Even Darwin, we now know from James Moore, finally abandoned Christianity because of theological questions raised by the deaths of his father and his ten-year-old daughter, Annie, not because of his growing attachment to evolution.[41]

By and large, the Catholic naturalists in the Academy remained Catholic, the Presbyterians remained Presbyterian, and the agnostics remained agnostic. Though the churchgoers in the group tended to remain loyal to their denominations and to a Christian world-view, their acceptance of evolution sometimes forced them, as Jon H. Roberts has pointed out, "to reassess the relationship between nature and the supernatural." For example, as Joseph LeConte moved from seeing species as being "*introduced by the miraculous interference of a personal intelligence*" to viewing them as the products of divinely ordained natural laws, he rejected all "anthropomorphic notions of Deity" for a God "ever-present, all-pervading, ever-acting." However, Darwinism generally exerted a "nonrevolutionary" effect on the religious lives of those who accepted it, a point A. Hunter Dupree made years ago.[42]

The National Academy of Sciences harbored a number of religious skeptics, but most of them judiciously refrained from attacking the cherished beliefs of others. A notable exception was Edward S. Morse, the outspoken son of a Baptist deacon who won notoriety in Christian circles for making evolution his religion. Unlike his more prudent colleagues, he boldly confronted what he saw as the radical theological implications of Darwinism. As early as 1860 he warned a friend that accepting Darwin's theory "would leave one without a God" because "the origin of species according to his idea would be simple chance and nothing else." Years later he scoffed at the efforts of liberal preachers such as Henry Ward Beecher to harmonize evolution and Christianity. Such naive reconcilers, he argued, failed to see that humanity's "origin from lower forms of life knocks in the head Adam and Eve, hence original sin, hence the necessity for vicarious atonement, hence everything that savors of the bad place." Yet Darwinism had not destroyed his religious faith. He had already traded "stiff-necked orthodoxy" for free thought and deism by the time the *Origin* fell into his hands.[43]

In an influential analysis of the Darwinian controversies James Moore has pointed to the frequency with which evolution precipitated "spiritual crises" in the lives of those forced to contend with it. Once again, however, the Academy naturalists provide scant support for such an interpretation, either because the men escaped such crises or because of their reticence to reveal such private matters. In partial confirmation of his thesis, Moore cites the alleged experiences of two Americans, James Dwight Dana and Jeffries Wyman, whom earlier scholars had described, respectively, as experiencing "a long soul-searching struggle" over evolution and as suffering from "deep distress, emotional as well as rational," over the prospect of apelike ancestors. Moore neglects, however, to mention that his authority for Dana pointedly stated that, despite his expectations, he had found "no evidence" to support the supposition that "an inner conflict involving his religious beliefs" lay behind Dana's struggle. And a recent study of Wyman, based on new evidence, has concluded that Wyman experienced "very little difficulty in embracing evolution." The most striking evidence of inner turmoil that I have found comes from Joseph LeConte, who graphically recounted having "sometimes wrestled in agony . . . with this demon of materialism." But even he linked evolution with materialism only indirectly, and he emerged from these intellectual struggles not only spiritually unscathed but, he thought, more theologically enlightened.[44]

Occasionally, historians have suspected a correlation between theological convictions and evolutionary beliefs. In a study of twenty-odd Victorians who wrote about Darwinism, Moore has claimed that "only those who maintained a distinctly orthodox theology," generally Calvinists, imbibed Darwinism undiluted, while liberal Christians, such as Unitarians, tended to adulterate Darwin's theory with Lamarckian and orthogenetic confections to make it more compatible with their belief in the inevitability of human progress.[45] The implied affinity between Darwinism and Calvinism is hard to sustain when we look at the naturalists in the Academy. Of the dozen who might have identified themselves as Calvinists (Presbyterians, Congregationalists, and Reformed communicants), Edward Hitchcock, George Engelmann, and Arnold Guyot resisted evolution; John Torrey avoided the issue; Lewis Henry Morgan excluded humans from the scheme; William Barton Rogers, Joseph LeConte, and Charles D. Walcott embraced catastrophism; Asa Gray fashioned his own theistic version of natural selection; and Benjamin Silliman, George H. Cook, and Sereno Watson left no known record of their views. With the four Roman Catholics—James Hall, Eugene W. Hilgard, Thomas Sterry Hunt, and Samuel S. Haldeman—the picture is no clearer. Haldeman paved the way for evolution, Hilgard advocated a paroxysmal view, Hunt expressed skepticism, and Hall refused to say anything. Among the Unitarians, Louis Agassiz and J. P. Lesley rejected evolution, while Joseph Leidy, Jeffries Wyman, and J. D. Whitney were among the first to accept it. If there is a pattern to these diverse responses, I fail to see it.[46]

Why Evolution?

Early in the Darwinian debates the American philosopher Chauncey Wright observed a striking "paradox" in the response of his contemporaries to Darwin: Increasing numbers were adopting the English naturalist's views on transmutation while at the same time rejecting natural selection, his explanation for it. It thus appeared that Darwin had "won a victory, not for himself, but for Lamarck." More recently Peter Bowler has addressed this "apparent paradox" in his provocatively titled book *The Non-Darwinian Revolution.* While readily conceding the rapid shift from belief in a special creation to belief in organic evolution in the years after the appearance of the *Origin,* he dismisses the so-called Darwinian revolution as "a historical myth."[47] To some, Bowler's revisionist views might seem extreme, but in one sense they accurately reflect the attitudes of America's leading naturalists. For, as we have seen, among the entire group of Academy evolutionists in the nineteenth century, not one in the 1860s and 1870s pushed natural selection as far as Darwin had done in 1859, and none came close to equaling Weismann in relying on natural selection in the 1880s and 1890s. The revolution sparked by the *Origin* overthrew special creation but failed to install natural selection in its place.

If American naturalists did not embrace organic evolution because Darwin offered them a plausible mechanism in natural selection, then why did they? The most compelling empirical evidence in support of evolution came to light in the early 1870s, when the American paleontologist O. C. Marsh uncovered the fossil remains of toothed birds in Cretaceous rocks from Kansas. Such discoveries, connecting ancient reptiles and birds, served, in Marsh's words, as "the stepping-stones by which the evolutionist of to-day leads the doubting brother across the shallow remnant of the gulf once thought impassable." Marsh's findings, raved the British zoologist Thomas H. Huxley, elevated Darwin's speculations about missing links "from the region of hypothesis to that of demonstrable fact." In 1876 Marsh won even greater fame for assembling a series of equine fossils that remarkably documented the evolution of the horse in North America. This evidence prompted Huxley, then lecturing in the United States, to announce that "the doctrine of evolution, at the present time, rests upon exactly as secure a foundation as the Copernican theory of the motions of the heavenly bodies did at the time of its promulgation." A few years later Darwin congratulated Marsh for providing "the best support to the theory of evolution, which has appeared within the last 20 years."[48] True enough, but Marsh's impressive illustrations of evolution became available too late to account for the shifting opinions of most Academy naturalists.

In a recent analysis of the dynamics of intellectual revolutions, including the one associated with Darwin, the historian of science Frank J. Sulloway attaches considerable significance to the participants' personalities, which, he argues, were formed in large part by the order in which they were born into their families and by the ensuing competitive strategies they adopted in dealing with their siblings. Because of the distinctive domestic niche they occupy, firstborns tend to be more conforming, conscientious, and assertive than their younger siblings, who are inclined to be less rigid and less fearful of new ideas and experiences that challenge the status quo. Thus radical intellectual revolutions have been led predominantly by laterborns, while firstborns, regardless of age, have been far more likely to oppose such changes. For 228 worldwide participants in the Darwinian revolution between 1859 and 1875, Sulloway shows that laterborns were 4.6 times as likely to support evolution as firstborns. Such findings have led him to reject the emphasis that Marxist historians of science have placed on social class in explaining the Darwinian revolution and to insist that "the focus of the battle over the theory of evolution was *within* the family, not *between* families." By combining information about birth order with information about age and social attitudes, two other significant predictors, Sulloway has been able to assign much of the responsibility for an individual scientist's position on Darwinism.[49]

At first glance the Academy naturalists seem not to conform to Sulloway's model, especially with respect to birth order. For example, Samuel Haldeman, the earliest of the Academy group to treat evolution sympathetically in a public forum, was an eldest child, as were Asa Gray and J. D. Whitney, two of the first three Academy naturalists to endorse Darwinism openly (the third being William Barton Rogers, a second son). E. D. Cope, another early convert to evolution, was also an eldest child. On the antievolutionary side of the debate, where one might expect to find largely firstborns, Arnold Guyot, Edward Hitchcock, J. P. Lesley, and John Newberry were all laterborns. Agassiz, the arch-critic of evolution, was indeed the eldest surviving child in his family, but the personality traits he acquired at home had not prevented him in the years prior to 1859 from promoting novel, indeed iconoclastic, ideas in geology and ethnology.

However, generalizing from the few examples given above would be misleading. A statistical analysis that includes four variables—birth order, social attitudes (including religious views), age, and social class—provides support for Sulloway's claims; see the table on page 46. The first three variables are indeed significant predictors of American responses to Darwinism, just as they are in Sulloway's model for the Darwinian revolution as a whole. By contrast, social class has no predictive significance.

What does this table tell us about the Academy naturalists? From the statistics presented there, we can determine that laterborns in the Academy were 1.8 times as likely to embrace Darwinism as firstborns (controlled for their greater numbers in the population being sampled). This odds ratio rises to 2.6 when it is stratified by social attitudes. Although the influences of birth order among the academicians are not as striking as those Sulloway obtained for his own smaller sample of American scientists (for whom the odds ratio is 5.0 in favor of laterborns), they are statistically significant—a result all the more impressive given the fact

The Significance of Four Variables in Explaining American Responses to Darwinism

Sample	Variable	Partial correlation	Bivariate correlation	t	p
1. Academy naturalists ($N = 80$)	Birth order	.27	.13	2.18	.04
	Social attitudes	.30	.45	2.47	.02
	Age	−.27	−.45	−2.18	.04
	Social class	.00	−.08	0.02	.98 (n.s.)
2. Sulloway's Americans ($N = 64$)	Birth order	.34	.23	2.67	.02
	Social attitudes	.53	.63	4.62	.0001
	Age	−.28	−.32	−2.17	.04
	Social class	.08	−.06	0.62	.54 (n.s.)
3. Combined samples 1 and 2 ($N = 116$)	Birth order	.32	.17	3.28	.001
	Social attitudes	.55	.56	6.33	.0001
	Age	−.24	−.33	−2.32	.03
	Social class	.15	.03	1.44	.16 (n.s.)

Note: This table gives the correlations between responses to Darwinism, rated by ten historical experts on a 7-point scale, and four variables: birth order (being laterborn), social attitudes (being liberal religiously and politically), age at which opinions on Darwinism were first expressed (being older), and social class (being lower). N identifies the number of individual participants (the number of participants in Samples 1 and 2 does not add up to the number in Sample 3 because some men appear in both groups). N.s. indicates that the correlations are not significant. All statistical results pertain to the partial correlations, and include p (which indicates the likelihood of the correlation happening by chance; e.g., $p < .05$, the conventional upper limit for being statistically significant, shows that the odds of a result occurring by chance are less than one in twenty) and two-tailed t values (a measure of effect size). For Sample 1 the degrees of freedom are 1/61 and the harmonic mean for the overall N is 66. For Sample 2 the respective figures are 1/56 and 61; for Sample 3, 1/92 and 97. Frank J. Sulloway generously calculated these statistics using BMDP's 8D Program (Correlations with Missing Data) and BMDP's 9R Program (All Possible Subsets Regression), thus allowing an exact comparison between his sample and mine. I owe him a significant ($p < .0001$) debt of gratitude.

that the younger Academy naturalists first expressed their opinions after the mid-1870s, by which time the scientific debate about the occurrence of evolution had all but ended, making birth order a largely irrelevant factor in determining individual positions. The combination of birth order with social attitudes and age (Sulloway's three strongest predictors of attitudes toward Darwinism) explains 26 percent of the variance among Academy naturalists, which is equivalent to classifying correctly the attitudes toward evolution of 75 percent of the naturalists. Among those who spoke out on Darwinism in all three samples, the ones who displayed liberal religious and political attitudes were significantly more likely to endorse Darwin's theories than those who did not. Age is also a significant predictor in all three samples, with younger participants being more supportive of Darwin's theories than older ones. The influence of social class is not statistically significant for the Academy naturalists, for Sulloway's American sample, or for the combined sample.

Few of the Academy naturalists ever spelled out their reasons for accepting evolution, but a number of them did comment on the factors that contributed to Darwin's celebrated success. Rather than emphasizing the novelty of his views, they commonly pointed to the correspondence of his teachings with beliefs already widely held and his skill in constructing a persuasive argument from familiar evidence. "The time was ripe" for a new theory, observed O. C. Marsh, because the old belief in "special creations" had already been "undermined by well established facts." Asa Gray made the same point, noting that "a notable proportion of the more active-minded naturalists had already come to doubt the received doctrine of the entire fixity of species, and still more that of their independent and supernatural origination." With so many readily available facts suggesting evolution, the only remaining need, explained Edward D. Cope, was for "a courageous officer to marshal them into line, a mighty host, conquering and to conquer."[50]

Even more compelling than the positive evidence for evolution was the overwhelmingly negative sentiment against special creation. As early as 1838 Darwin had concluded that attributing the structure of animals to "the *will* of the Deity" was "no explanation—it *has not the character of a physical* law & is therefore utterly useless." As Asa Gray put it, the great strength of evolution appeared "on comparing it with the rival hypothesis . . . of immediate creation, which neither explains nor pretends to explain any [facts]." By 1859 a growing number of American naturalists, like Gray, had become self-conscious, even embarrassed, about having to resort to supernatural interventions when seeking to explain the origin of species.[51]

The understandable reluctance of creationists to translate their convictions into scientific language made them the objects of derision. The ichthyologist Theodore N. Gill, who complained about the "vague and evasive" responses that nonevolutionists gave to inquiries about the processes of creation, quoted Darwin in demanding answers to such questions as: "Did 'elemental atoms flash into living tissues?' Was there vacant space one moment and an elephant apparent the next? or did a laborious God mould out of gathered earth a body to then endue with life?" In Gill's opinion, such information was a prerequisite to conceiving of creation in any scientifically useful way.[52]

Given the scientific vacuity of creationist beliefs, the overwhelming attraction of evolution, as one layman bluntly phrased it, was that there was "literally nothing deserving the name of Science to put in its place." The young geologist William North Rice, a student of Dana's, made much the same point even before fully embracing evolution. "The great strength of the Darwinian theory," he wrote in 1867, "lies in its coincidence with the general spirit and tendency of science. It is the aim of science to narrow the domain of the supernatural, by bringing all phenomena within the scope of natural laws and secondary causes."[53]

In reviewing Darwin's *Origin of Species* for the *Atlantic Monthly,* Asa Gray addressed the question of how he and his colleagues had come to feel so uncomfortable with a "supernatural" account of speciation. "Sufficient answer," he explained, "may be found in the activity of the human intellect, 'the delirious yet divine desire to know,' stimulated as it has been by its own success in unveiling the laws and processes of inorganic Nature." Minds that had witnessed the dramatic progress of the physical sciences in recent years simply could not "be expected to let the old belief about species pass unquestioned." Organic evolution, echoed Gray's friend George Frederick Wright, accorded with the fundamental principle of science, which states that "we are to press known secondary causes as far as they will go in explanation of facts. We are not to resort to an unknown (i.e., supernatural) cause for explanation of phenomena till the power of known causes has been exhausted. If we cease to observe this rule there is an end to all science and all sound sense." Other commentators made the same point.[54]

During the decades, even centuries, leading up to the appearance of the *Origin* the advance of natural law had come to represent the very essence of scientific progress. By 1859 the great majority of American naturalists, whatever their philosophical stripe, agreed that miracles no longer had a legitimate place in science. They rapidly and successfully negotiated the potentially hazardous transition from special creation to evolution not because of Darwin's particular theory of origins but because of their nearly universal commitment to methodological naturalism, a trait common to Darwinians and Lamarckians, neo-Darwinians and neo-Lamarckians alike. The publication of the *Origin* thus served more as the catalyst for the shift to organic evolution than as its cause.[55]

2

CREATING CREATIONISM: MEANINGS AND USES SINCE THE AGE OF AGASSIZ

In an elegant essay titled "Deconstructing Darwinism" James Moore several years ago noted how strikingly "little effort had been expended by historians of science in tracing the proliferation of Darwin-related vocabulary and interpreting its function in public discourse." To help remedy this deficiency, he carefully examined the metamorphosis of the term "Darwinism" from a synonym for naturalistic evolution generally to a label for evolution by natural selection specifically, illustrating in the process how malleable and politically serviceable such labels can be. Recently Mark A. Noll and David N. Livingstone drew attention to the historical significance of efforts to define terms in the Darwinian debates. Definitions "are designed to demarcate the true from the false, the legitimate from the illegitimate, the relevant from the irrelevant," they wrote. "Accordingly, the control of definitions is of enormous consequence for intellectual debate, since definitions position both ideas and people on particular sides of debates." Like Moore, they focused on Darwinism.[1]

Despite the utility of such exercises, no one has yet looked at the deployment of the terms "creationism" and "creationist" in the late nineteenth century and the twentieth. Even I, in a 450-page book on the history of modern creationism, failed to address the issue of when these terms first came into use.[2] In this chapter I will draw on my previous research and some subsequent explorations to address the following queries: When, why, and how were the terms "creationism" and "creationist" first employed? At what point did they come to be associated primarily with the community of evangelical Christians? And under what circumstances did those militant evangelicals known as Fundamentalists capture creationism to represent their own distinctive interpretation of Genesis? Most of my discussion will concentrate on American usages, from the Darwinian debates of the late nineteenth century through the Fundamentalist controversy of the early twentieth century and on to the rise of "scientific creationism" in recent decades.

Creation and Evolution in the Nineteenth Century

When Charles Darwin published *On the Origin of Species* in 1859, the term "creationist" commonly designated a person who believed in the special creation of a soul for each human fetus, as opposed to a traducianist, who believed that the souls of children were inherited from their parents. For example, when the Princeton theologian Charles Hodge devoted a section of his *Systematic Theology* to "creationism," he limited his discussion entirely to the origin of the soul. Nevertheless, just one day after the *Origin of Species* appeared, Darwin employed the creationist label to refer to opponents of evolution. "What a joke it would be," he wrote to Thomas Henry Huxley, "if I pat you on the back when you attack some immovable creationist!" Since at least the early 1840s Darwin had occasionally referred to "creationists" in his unpublished writings, but the epithet acquired little public currency. In 1873 Asa Gray described a "special creationist" (a phrase he placed in quotation marks) as one who maintained that species "were supernaturally originated just as they are," and in 1880 he briefly contrasted Darwinism with "direct Creationism." Similarly, in the 1890s the priest-scientist John A. Zahm of Notre Dame occasionally used "creationist" as a synonym for antievolutionist. But the practice of describing antievolutionists as creationists remained relatively infrequent during the nineteenth century—and confined, it seems, largely to people who no longer believed in special creations. As far as I can tell, such prominent North American anti-Darwinists as Louis Agassiz, Arnold Guyot, and John William Dawson neither called themselves creationists nor referred to their views as creationism. During the seventy-five years or so after the appearance of the *Origin of Species* opponents of evolution were commonly denoted by such terms as "advocates of creation," "exponents of the theory of the immutability of species" or, increasingly, "antievolutionists."[3]

One of the reasons why the label "creationist" languished in obscurity for so long was the great diversity of opinion among the dissenters from

Darwinism. In the *Origin of Species* Darwin sought to discredit "the ordinary view of creation," but, as we saw in Chapter 1, he failed to identify the particular view he had in mind. Almost certainly he was not referring to the popular idea, associated with both the author of Genesis and the eighteenth-century Swedish naturalist Carolus Linnaeus, that the originally created pairs of animals and humans had dispersed from a central point, such as the biblical Garden of Eden, to populate Earth. More likely he had in mind the hypothesis of his friend Charles Lyell that there were various "*centres* or *foci* of creation," where new plants and animals appeared when needed—or possibly Louis Agassiz's notion of species being created in large numbers to repopulate Earth following global catastrophes. Agassiz's colleague Asa Gray left no doubt in his writings that he considered Agassiz's view "the commonly received doctrine" of creation.[4]

It is impossible to determine how many Americans followed Agassiz in postulating multiple, plenary creations. Most conservative Christians never expressed themselves on the matter, though many of them undoubtedly clung to the traditional view that God had created the heaven and the earth in six literal days about 6,000 years ago. This reading of Genesis necessitated rejecting the rapidly growing body of geological and paleontological evidence of Earth's great antiquity, but it could be defended on the grounds that scientists had not correctly interpreted their data. At mid-century one American observer estimated that perhaps "one half of the Christian public" still adhered to this position. The remainder, he claimed, had divided largely into two rival camps: those who accommodated the findings of historical geology by interpreting the days of Genesis 1 to represent vast ages in the history of the earth (the day-age theory) and those who did so by separating a creation "in the beginning" from a much later Edenic creation in six twenty-four-hour days (the gap theory).[5]

To compound the confusion about special creation, its most vocal scientific proponents disagreed markedly over the number of supernatural interventions required. As we have seen, Agassiz, a fair-weather Unitarian, called for a virtual infinitude of miracles, while his Swiss-born compatriot Arnold Guyot, an evangelical Presbyterian, demanded only three divine intrusions into the natural order: for the special creation of matter, life, and humans. The Presbyterian geologist John William Dawson, who served as principal of McGill University in Montreal, took an intermediate position. He followed his mentor Charles Lyell in postulating successive creations in various "centres" but stopped short of insisting that "all groups of individual animals, which naturalists may call species, have been separate products of creation."[6]

Although a few advocates of creation, such as the British naturalist Philip Gosse, indulged in speculating about the details of creation—Gosse suggested, for example, that God had created Adam with a navel, thus giving him the appearance of age at the time of his creation—most non-evolutionists resolutely avoided the topic. Louis Agassiz on more than one occasion frustrated colleagues by refusing to provide a single specific description of how a species came into existence. "When a mammal was created, did the oxygen, hydrogen, nitrogen, and carbon of the air, and the lime, soda, phosphorus, potash, water, &c., from the earth, come together and on the instant combine into a completely formed horse, lion, elephant, or other animal?" inquired Agassiz's Harvard colleague Jeffries Wyman. If this question is "answered in the affirmative, it will be easily seen that the answer is entirely opposed by the observed analogies of nature."[7]

Creationism and Its Critics

In 1899 Dawson, the last major nineteenth-century scientist to defend special creation, died. By this time the scientific community was promoting organic evolution as an established "fact," even as members quarreled among themselves over the exact mechanisms of change. About the only Americans left debating the merits of special creation were conservative, often evangelical, Christians. In the early 1920s the most concerned critics of human evolution launched a movement to eradicate the offending belief from the churches and schools of America. But throughout the so-called Fundamentalist controversy, their goal remained the elimination of evolution, not the promotion of a particular doctrine of creation. They dedicated their organizations to "Christian Fundamentals," "Anti-Evolution," and "Anti-False Science," not to creationism.[8]

Evangelicals had still reached no consensus about the correct reading of Genesis 1, although even the most conservative commentators had come to terms with the antiquity of life on Earth and a deluge of local or geologically superficial significance. Leaders of the Fundamentalist movement tended to promote either the day-age theory, endorsed by William Jennings Bryan and William Bell Riley, or the gap (or ruin-and-restoration) interpretation, taught in the popular *Scofield Reference Bible* and preached by the evangelist Harry Rimmer. About the only Christians to insist on the recent appearance of life and on a fossil-burying flood were the Seventh-day Adventist disciples of Ellen G. White, who claimed to have witnessed the creation of the world in vision (see Chapter 5). Shortly after the turn of the century the self-instructed Adventist geologist George McCready Price began advocating a scientific version of White's views that he called "the new catastrophism," "the new geology," or simply "flood geology." According to Price, a correct reading of Genesis 1 ruled out any notion of "creation on the installment plan"—creative acts interspersed over millions of years—which he regarded as a "burlesque" of creation. The gap theory required too much verbal "dodging and twisting" to conform to his standards of biblical literalism, and the day-age theory was even more egregious. This "libel on Moses" struck at the very basis of the Sabbath—and the identity of Seventh-day Adventists—by suggesting that the seventh day of creation had not been a literal twenty-four-hour period. The Bible, as Price saw it, allowed for only *"one act of creation,"* which, he said, "may easily be supposed to have included all of those ancestral types from which our modern varieties of plants and animals have been derived."[9]

Early in his career Price toyed with the idea of starting a magazine called *The Creationist* to popularize his views, but for years he avoided equating his theory of flood geology with creationism generally. As far as I can tell, it was not until 1929 that one of his students, Harold W. Clark, a Berkeley-trained biologist, explicitly packaged Price's new catastrophism as "creationism." In a brief self-published book titled *Back to Creationism* Clark urged readers to quit simply opposing evolution and to adopt the new "science of creationism," by which he clearly meant Price's flood geology.[10]

In the mid-1930s Dudley Joseph Whitney, one of Price's earliest non-Adventist converts to flood geology, observed that Fundamentalists were "all mixed up between geological ages, Flood geology and ruin, believing all at once, endorsing all at once." How, he wondered, could evangelical Christians possibly convert the world to creationism if they themselves could not even agree on the meaning of Genesis 1? To bring order to this chaotic state of affairs, he pushed for the establishment of a society to create a united Fundamentalist front against evolution, the Religion and Science Association. In a press release Price declared that the association backed flood geology as "by far the best and most reasonable explanation of the facts of the fossils and the rocks," while "condemning and repudiating the only other possible alternatives about the fossils: (a) The Day-Age theory, which is false scientifically and cannot be made to harmonize with the record of Genesis I; and (b) The Pre-Adamic Ruin Theory, which makes nonsense of the scientific facts and is utterly fantastic theologically." This description represented mere wishful thinking on Price's part. The infant association died within two years of birth, largely because its leaders could not agree on a common interpretation of the first chapter of Genesis.[11]

Unfortunately for Price and Whitney, they had recruited for the association's presidency a Wheaton College chemist, L. Allen Higley, who, like so many Fundamentalists, believed in the truth of the ruin-and-restoration theory as adamantly as they believed in flood geology. In Whitney's opinion, the views Higley espoused had become "a *Fundamentalist* dogma; and when I say dogma in that connection I mean DOGMA and then some. . . . The thing is an *obsession.*" Whitney had initially hoped to convert the "Wheaton crowd" to flood geology, believing that "the more Wheaton can be played up in our association, the more influence it will have with the Fundamentalist." When the Wheaton establishment proved immovable, Whitney despaired of the future for flood geology. Nevertheless, in 1937 he began circulating a mimeographed sheet, *The Creationist,* hoping thereby to identify creationism with the Price school.[12]

Such efforts to co-opt the creationist label did not go uncontested for long, in part because evangelical scientists who did not accept flood geology could ill afford to have their constituencies view them as noncreationists. Thus leaders of the American Scientific Affiliation, organized in 1941, waged a two-pronged campaign in the 1940s and 1950s to discredit flood geology as "pseudo-science" while cloaking their own quasi-evolutionary views in the mantle of creationism. The anthropologist James O. Buswell III, who introduced fellow evangelicals to the evidence for human antiquity and development, decried the common tendency of evolutionists to equate creationists with "hypertraditionalist" Fundamentalists such as Price. He winced when he heard the paleontologist George Gaylord Simpson declare in 1950 that "creationists are found today only in non- or anti-scientific circles." Buswell argued for a broad understanding of creationism that would allow both rigid hypertraditionalists and progressive scientists such as himself, "who constantly allow their interpretations to be open to the acceptance of newly discovered facts," to be called creationists. He boldly, but unsuccessfully, urged evangelicals ranging from progressive creationists to theistic evolutionists to march under the banner of "scientific creationism."[13]

The Wheaton biologist Russell L. Mixter, who gently prodded the ASA to accept more and more evidence of organic evolution, pushed for a similarly inclusive definition of creationism. "Creationists of today are not in agreement concerning what was created according to Genesis," he announced in a controversial mid-century monograph in which he defended the evolution of species within major groups of animals. "In this sense," he noted provocatively, "Creationists can be called evolutionists." Indeed, they could; but as he soon discovered, they might not want to be if they desired to teach at schools such as Wheaton. In terms of job security at Christian colleges, it was far better to be an evolutionist who called himself a creationist. Adopting Mixter's nomenclature, Wheaton president V. R. Edman was able to assure a concerned board of trustees in 1957 that "we at Wheaton are avowed and committed creationists."[14]

The ASA progressives received a boost in 1954, when Bernard Ramm, a Baptist philosopher and theologian, brought out an influential treatise called *The Christian View of Science and Scripture.* Convinced that Price's scientifically disreputable views had come to form "the backbone of much of Fundamentalist thought about geology, creation, and the flood," he urged fellow evangelicals to repudiate such "narrow bibliolatry" and adopt what he called "progressive creationism," a position far to the left of fiat creationism but slightly to the right of theistic evolution. It did away with the necessity of believing in a young Earth, a universal flood, and the recent appearance of humans, but still required that "from time to time the great creative acts, *de novo,*" had taken place.[15]

Such attempts to stretch the meaning of creationism—almost to the point of accepting divinely guided evolution—provoked a backlash among the increasingly outspoken advocates of Price's flood geology. In 1961 John C. Whitcomb, Jr., and Henry M. Morris published *The Genesis Flood,* which sought to establish a recent special creation and flood geology as the only orthodox understanding of Genesis. Two years later a group of ten like-minded scientists formed the Creation Research Society to promote this point of view. To enable their views to be taught in the public schools of America, these creationists in the early 1970s peeled off the biblical wrappings of flood geology and repackaged it as "creation science" or "scientific creationism." This relabeling reflected more than euphemistic preference; it signified a major tactical shift among strict six-day creationists. Instead of denying evolution its scientific credentials, as biblical creationists had done for a century, these scientific creationists argued for granting creation and evolution equal scientific standing. And instead of trying to bar evolution from the classroom, as their predecessors had done in the 1920s, they fought to bring creation into the schoolhouse and shunned the epithet "antievolutionist." "Creationism is on the way back," Morris proudly announced in 1974, "this time not primarily as a religious belief, but as an alternative scientific explanation of the world in which we live."[16]

The ASA's drift toward theistic evolution and the CRS's insistence on young-Earth creationism made many self-described creationists feel left out in the cold. As the Canadian antievolutionist John R. Howitt, a gap theorist, put it, while many ASA members were embracing theistic evolution "or some such rubbish, the true fundamentalists [were] all going over to flood geology. Oh me, oh my!" In the mid-1960s he begged CRS leaders not to make flood geology a test of creationist orthodoxy. "Let it be truly a Creation Research Society," he urged, "seeking for truth, no matter where it ends up,—as flood geology, the gap theory, theistic evolution or what not." By the early 1970s, however, he had abandoned any hope of reconciliation. "The flooders are getting pretty dogmatic these days," he noted with sadness, and not a little irritation. Before long even in Great Britain, where evangelical antievolutionists had long kept the flood geologists at bay, the so-called Young Earthers took control of the Evolution Protest Movement, condemned the gap and day-age theories as unscriptural, and reinvented themselves as the Creation Science Movement.[17]

Although many evangelicals who regarded themselves as creationists, from Billy Graham to Jimmy Lee Swaggart, resisted the allure of scientific creationism, by the last decades of the twentieth century Price's intellectual heirs had virtually co-opted the creationist label for their own interests. Even their severest critics often conceded as much. When in 1984 the National Academy of Sciences issued an official condemnation of "creationism," that august body defined it as comprising beliefs in a young Earth and universe, flood geology, and the miraculous origination of all living things. Writing in the early 1980s in defense of the day-age theory and against flood geology, the Calvin College geologist Davis A. Young noted regretfully that, although he still believed in the biblical story of creation, he was opposed to creationism. This ironic turn of events had resulted because "those who advocate the creation of the world in seven literal days only a few thousand years ago have come to be known generally as creationists." When he and two Calvin colleagues, Howard J. Van Till and Clarence Menninga, later collaborated on a book titled *Science Held Hostage* (1988), they assigned equal blame for this terrorist act to "naturalism" and "creationism," which they explicitly identified with the views of the scientific creationists.[18] Latter-day flood geologists may not have liked being lumped together with godless evolutionists as enemies of true science (and religion), but they could only have appreciated the often grudging but increasingly widespread recognition that their once marginal views, inspired by the visions of an obscure Adventist prophetess, now defined the very essence of creationism.

3

DARWINISM IN THE AMERICAN SOUTH: FROM THE EARLY 1860S TO THE LATE 1920S

No region in the world has won greater notoriety for its hostility to Darwinism than the American South. Despite the absence of any systematic study of evolution in the region, historians have insisted that southerners were uniquely resistant to evolutionary ideas. Rarely looking beyond the dismissals of Alexander Winchell from Vanderbilt University in the 1870s and James Woodrow from Columbia Theological Seminary in the 1880s—or the Scopes trial in the 1920s—they have concluded, in the words of Monroe Lee Billington, that "Darwinism as an intellectual movement . . . bypassed Southerners." W. J. Cash, in his immensely influential book *The Mind of the South,* contended that "the overwhelming body of Southern schools either so frowned on [Darwinism] for itself or lived in such terror of popular opinion that possible heretics could not get into their faculties at all or were intimidated into keeping silent by the odds against them." Darwin's few southern converts either "took the way of discretion" by moving to northern universities or so qualified their discussions of evolution as to render the theory "almost sterile."[1]

Historians of religion and of science have generally concurred with the judgment of southern historians. Uncompromising antievolutionism, says the American church historian George M. Marsden, "seems more characteristic of the United States than of other countries and more characteristic of the South than of the rest of the nation." Because people in the region were more religiously conservative and less well educated than people in the North, such differences were only to be expected. The historian of

This chapter was written with Lester D. Stephens.

science David N. Livingstone echoes Marsden, describing antievolutionism as "mainly a southern phenomenon."[2]

I have no desire to discount southern resistance to organic evolution. The evidence for that is ample—and no doubt in greater quantities than for other regions in the United States. Jon H. Roberts is probably correct when he suggests in *Darwinism and the Divine in America* that "although ardent Biblicists could be found in every geographical region in the United States, a slightly disproportionate number of them resided in the southern and border states." Besides, southern intellectuals undoubtedly accepted evolution more slowly than did their northern counterparts. Unlike many northerners who eagerly embraced Darwinism, observed A. T. Robertson in 1885, "the more plodding scholars of the South advanced slowly, seeing well their way, and being firmly convinced of the reality of the claims of evolution before accepting it."[3] My point is that the South was far less uniform in its opposition to Darwinism than most scholarly accounts suggest. In fact, the very success of Darwinism in the South contributed significantly to the outburst of antievolutionism in the 1920s. Although far from definitive, my survey of southern responses to evolution, conducted with Lester D. Stephens, shows that conventional wisdom about the controversies associated with Winchell, Woodrow, and Scopes give a highly distorted picture of southern attitudes toward evolution in the years from the early 1860s to the late 1920s.

Genesis and Geology before the *Origin*

The South's reputation for being inhospitable to science dates back to the pre-Darwinian period, when, according to one eminent historian, "the cotton kingdom . . . killed practically every germ of creative thought." Certainly the antebellum South's overall record for scientific achievement was no match for the Northeast's, but in urban centers such as Charleston and New Orleans and in college towns scientific activity often flourished. In the decades before the American Civil War the slave states of the South and border regions supported as many leading scientists per 100,000 white residents as the free states to the north and west.[4] And in responding to scientific developments of concern to many Christians—such as the nebular origin of the solar system, the antiquity of Earth, and the plural creation of human races—southerners could be found on both sides of the debate.[5]

When the young American Association for the Advancement of Science met in Charleston in 1850, local fears about maintaining the harmony between Genesis and geology prompted the northern visitor Alexander Dallas Bache to comment that the same wave of opposition to geology that had swept over the North twenty years earlier was now washing over the South. Perhaps there was some justification for Bache's observation, but southern intellectuals were not nearly as hostile to geology as his reaction suggests. Even the hide-bound Presbyterian theologian Robert Lewis Dabney, one of the region's most vocal critics of geological efforts to stretch the history of Earth beyond Eden, stopped short of damning Christian geologists who sought to harmonize the Bible with the testimony of the rocks. He did, however, express resentment over scientists' "continual encroachments . . . upon Scripture teachings." At first they had requested only a pre-Adamite Earth; now they were demanding acceptance of the nebular hypothesis, a local flood, ancient humans, and even organic evolution.[6]

Numerous southerners in the years before 1859 openly pushed for reinterpreting Genesis in the light of modern geology—and apparently suffered few ill effects for their boldness. Michael Tuomey, for example, in his 1848 *Report on the Geology of South Carolina,* sought to accommodate the findings of geology by inserting an immense span of time between the creation "in the beginning" and the much later Edenic creation. Similarly, the paleontologist and physician Robert W. Gibbes assured members of the South Carolina Institute that Mosaic silence on the date of the original creation permitted Christians to accept the notion that "the earth has been inhabited by animals and adorned with plants during immeasurable cycles of time antecedent to the creation of man." Richard T. Brumby of South Carolina College also took this position. At first Brumby felt intellectually isolated from friends who clung to the doctrine of a recently "finished" earth, created in six literal days, but before his death in the mid-1870s he rejoiced to see that that teaching had finally been "exploded." Indeed, by that time it seemed that "most intelligent Christians" agreed that Adam and Eve had arrived on Earth long after other members of the animal kingdom.[7]

Southern Presbyterians and the Woodrow Affair

In his study of the "gentlemen theologians" of the old South, E. Brooks Holifield found that these literate ministers generally displayed great enthusiasm, not hostility, toward natural science; yet he detected "signs of strain" beginning to appear in the 1850s. Indicative of the growing ambivalence of some southerners toward science on the eve of the Civil War was the creation of the Perkins Professorship of Natural Science in Connexion with Revealed Religion, established at the Presbyterian Columbia Theological Seminary in South Carolina. In 1857 the Tombeckbee Presbytery in Mississippi had unanimously adopted a resolution calling for a professorship on science and religion "to forearm and equip the young theologian to meet promptly the attacks of infidelity made through the medium of the natural sciences." The Presbyterians of Mississippi had grown alarmed by recent scientific theories, such as those espoused by the still unknown author of *Vestiges of the Natural History of Creation*, and they worried that an inexperienced minister, posted to a backwoods parish, might have his faith tested by some impertinent infidel, perhaps a physician who had simultaneously learned how to heal the body and "kill the soul." But in addition to protecting young preachers from such attacks, the Tombeckbee Presbyterians wanted to provide seminarians "with such enlarged views of science, and its relationship to revealed religion, as will prevent them from acting with indiscreet zeal in defending the Bible against the supposed assaults of true science."[8]

The Columbia seminary selected as the first occupant of the chair an English-born scientist-cleric, James Woodrow, a former student of Louis Agassiz's at Harvard who had gone on to the University of Heidelberg for a doctorate, presumably in chemistry. While teaching at the Presbyterian Oglethorpe College after returning from Europe, he had been ordained a Presbyterian minister. In 1861, the year he assumed the Perkins chair, he took over the editorship of the *Southern Presbyterian Review*, the leading quarterly of southern Presbyterianism, and four years later he added the weekly *Southern Presbyterian* to his editorial responsibilities. The latter magazine, he sometimes said, gave him a constituency of 4,000 readers. From 1869 to 1872 and again from 1880 to 1897 he also served on the faculty of the University of South Carolina.[9]

Woodrow, a social and theological conservative who professed to believe in "the absolute inerrancy" of the Bible, at first adopted a cautious stance. In his inaugural address as Perkins Professor he affirmed his acceptance of the antiquity of Earth and allowed that the deluge might have been a local affair, but he left no doubt about his rejection of the plurality of the human races. For twenty-four years he taught that evolution "probably was not true," but that "even if true it did not contradict or in any way affect the truth of the Scriptures." However, in preparing for a requested address to the seminary's alumni association in 1884, he reviewed the evidence in favor of evolution and became convinced that it was "probably true."[10] "We cannot go back to the beginning," he told the assembled alumni,

> but we can go a long way. The outline thus obtained shows us that all the earlier organic beings in existence, through an immense period, as proved by an immense thickness of layers resting on each other, were of lower forms, with not one as high or of as complex an organization as the fish. Then the fish appeared, and remained for a long time the highest being on the earth. Then followed at long intervals the amphibian, or frog-like animal, the reptile, the lowest mammalian, then gradually the higher and higher, until at length appeared man, the head and crown of creation.

Although Woodrow had come to believe that divinely guided evolution had produced man's body, he insisted that his soul had been "immediately created." And because of "insurmountable obstacles" connected with the biblical story of Eve, he continued to hold, as a biblical inerrantist, that both the body and the soul of the first woman had been specially created, a concession derided by antievolutionist critics as "unscientific." To harmonize his evolutionary views with the Mosaic account of creation, he adopted the day-age reading of Genesis 1 formulated by the Princeton geographer Arnold Guyot.[11]

The publication of Woodrow's address in both the *Southern Presbyterian Review* and in pamphlet form touched off a controversy that raged within southern Presbyterianism for four years. "At once a vehement attack upon him was begun," reported the Woodrow-edited *Southern Presbyterian,* "not apparently for his own views as given in the address, but on account of . . . the whole brood of Evolutionists from the beginning, especially the atheistic part of it, most of his assailants seeming not to have read the address at all." Leading the charge against Woodrow was his own colleague at the seminary, the powerful theologian John Lafayette Girardeau, who feared that enrollments would decline if Columbia became known as the "Evolution Seminary." In view of the recent closing of the institution for two years because of a shrinking student body, Girardeau's fears were not unfounded. Besides, at least one financial backer of the seminary had complained that "the Church did not give you money to have Darwinism taught."[12]

It quickly became clear that most southern Presbyterians had little use for "tadpole theology." Yet Woodrow was not without support. When the seminary trustees met in September 1884 to respond to the developing controversy, they voted eight to three to back Woodrow on the grounds that "the Scriptures, while full and clear in asserting the fact of creation, are silent as to its mode." This decision inflamed Woodrow's opponents, who quickly succeeded in reconstituting the board of trustees and obtaining a call for Woodrow's resignation. When he declined, the board fired him. Two of Woodrow's friends on the faculty quit in support of their beleaguered colleague, leaving Girardeau and his associates free, as the *Southern Presbyterian* put it, to form "a new 'Anti-Evolution Seminary.'"[13]

For several more years, however, the evolution question continued to preoccupy "the upper circles" of southern Presbyterians. At times it seemed as though the church was devoting "more zeal and attention to discussing the origin of Adam's body than to the interest of the souls of Adam's descendants." In 1886 alone, three different levels of church courts wrestled with it: the Presbytery of Augusta, which tried Woodrow for heresy; the four synods responsible for the seminary; and the General Assembly of the Presbyterian Church in the United States, which had received official complaints from no fewer than eight different presbyteries scattered throughout the South. By a vote of fourteen to nine, Woodrow won acquittal in Augusta, but the General Assembly adopted a hard-line position against human evolution:

> The Church remains at this time sincerely convinced that the Scriptures, as truly and authoritatively expounded in our confession of Faith and Catechisms, teach that Adam and Eve were created, body and soul, by immediate acts of Almighty power, thereby preserving a perfect race unity.
>
> That Adam's body was directly fashioned by Almighty God, without any natural animal parentage of any kind, out of matter previously created of nothing.
>
> And that any doctrine at variance therewith is a dangerous error. . . .[14]

But even this declaration did not bring the "long and violent warfare on Dr. Woodrow" to an end. His enemies appealed his acquittal to the

Synod of Georgia, which overturned the lower ecclesiastical court in a decision upheld in 1888 by the General Assembly. Although Woodrow took some pride in his role as a latter-day Galileo, and at times seemed to find exhilaration in combat, he anguished over having not only his orthodoxy but also his spirituality brought into question. At times he was forced to defend himself against charges ranging from "neglecting the sanctuary" (a result of poor personal and family health) to living an "almost wholly secularized" life (because of his involvement with numerous businesses). For years he fought to save his job and his reputation, but in the end even his friends conceded that "our little company [was] completely routed." Worse yet, at a time when John William Draper's best-selling *History of the Conflict between Religion and Science* (1874) was still on everyone's mind, the church's treatment of Woodrow had put "another javelin . . . in the hands of future John W. Drapers to hurl at the Church."[15]

At first glance the Woodrow affair seems strikingly to confirm accusations of southern antipathy to evolution. But a closer examination of this episode reveals the need to qualify this assessment. A majority of southern Presbyterians may have condemned Woodrow and his beliefs, but the embattled professor did win two key decisions, and over a hundred sympathizers at one time or another voted for him in the church trials. One Presbyterian pastor estimated that the pro-Woodrow faction constituted "at least one-tenth of our Church." Many of Woodrow's defenders would no doubt have described themselves as anti-Darwinists, but they did support his right to advocate a theistic version of evolution. At the height of the controversy an elderly alumnus of the Columbia seminary estimated that a "strong minority" of southern Presbyterians regarded evolution "as a mere scientific deduction, which may be true or not, but which can in no way imperil the interest of true religion, inasmuch as the Bible teaches the simple fact that God made Adam, but does not explain the mode or the particular substance out of which his body was formed." This same minister surmised that there were "scores of intelligent Christian men and women in almost all parts of the Church, who after studying the subject in the light of the discussions that have been going on for more than a year past have come to the conclusion . . . that Adam's body was 'probably evolved' from organic matter."[16]

It is also important not to generalize about southern attitudes toward evolution from the Woodrow controversy over human evolution. "Neither party," declared one participant, "denies that descent with modification is probably the law of the successive appearances of the *animal* tribes on this globe from the beginning until we come down to man. . . . We differ only upon one point, viz., the creation of the body of Adam." Another writer noted the same tendency to distinguish between animal and human evolution: "The point of discussion is . . . not Evolution in general. For life below man this is conceded generally, and one newspaper pronounces it 'harmless.' The controversy begins when the doctrine is applied to man."[17]

Because the opinions quoted above appeared in Woodrow's *Southern Presbyterian,* one might suspect that the writers were painting an overly positive picture of toleration for prehuman evolution in order to make all evolution seem less threatening. But the same distinction appears in the writings and speeches of the anti-Woodrow George D. Armstrong, southern Presbyterianism's leading voice on matters of science and religion. As a sometime science professor turned minister, he claimed to reject all forms of evolution "on grounds purely scientific." Nevertheless, he readily conceded that if evolution excluded the transition from inorganic to organic, at the beginning of the process, and from animals to humans, at the end, it was neither atheistic nor irreconcilable "with the Bible account of the origin of plants and animals in the world." Speaking before the General Assembly in 1886, he warned Christian evolutionists in the audience not to let evolution carry them "to the belief that it refers to man made in the image of God. It will necessitate giving up the doctrine of the fall." To ripples of laughter, he explained that "according to evolution, man was at his lowest stage, just evolved from a brute—how could he fall? he was already low as he could get."[18]

Southern Presbyterians were not alone in distinguishing between animal and human evolution. As they sometimes noted, even such leading northern lights as Princeton Theological Seminary's James McCosh, Francis Patton, and Archibald Hodge made the same distinction. "About the lower animals," explained Hodge, apparently speaking for his colleagues as well, "we are willing to leave it to the scientists as outside of immediate theological or religious interest."[19] In light of this widespread feeling, we should not assume in the absence of direct evidence that Woodrow's southern critics opposed all forms of organic evolution. And before we take Woodrow's dismissal from the faculty of a theological seminary as representative of southern intolerance of Darwinism, we should keep in mind that Woodrow, as the South's most notorious evolutionist, continued to serve undisturbed on the faculty of the University of South Carolina until his retirement in 1897, spending his last six years there as president.

In trying to account for why antievolution became "a standard test of the faith among southern evangelicals earlier than it did among northern fundamentalists," George M. Marsden draws on the Woodrow case to suggest that "the most likely principal explanation was that their northern counterparts had been infected by a liberal spirit, evidenced in the first instance in their unbiblical attacks upon slavery." Although this argument may have some merit, it finds little support in the actual debates over Darwinism. Southern Presbyterians were well aware that their opposition to evolution distinguished them from many northern brothers and sisters. Some feared that the northern church would call them "heretics" if they did not condemn evolution. Others, such as one of Woodrow's female correspondents, grieved "that the Northern Church should have occasion to comment upon a want of union among ourselves, with some little unchristian exultation."[20]

Indeed, even the Presbyterian press in the North seemed to relish reporting the monkey business disturbing fellow believers in the South. The *Presbyterian Journal* of Philadelphia, for example, called the attack on Woodrow "the ecclesiastical blunder of this generation" and accused his persecutors of yielding "to a spasm of terror." The *Interior,* a Presbyterian paper in Chicago, wondered editorially if there was "ever in the world such a thundering fiasco as the Woodrow business in the Southern Presbyterian Church!" Invoking meteorological metaphors, the editor concluded that "Southern cyclones do not have the faculty of catching on." Northerners told themselves that the internecine struggle in the South could never occur in their region. But despite the intense and at times acrimonious interregional rivalry, noted one of Woodrow's staunchest supporters, the northern church's *"toleration of Evolution has never been named"* in the long list of errors charged against it.[21]

Adamites and Pre-Adamites

Woodrow's fate at the Columbia seminary tells us as little about the overall reception of Darwinism in southern institutions of higher learning as does the equally celebrated expulsion of Alexander Winchell from the Methodist Vanderbilt University in Nashville, Tennessee. Winchell, a respected geologist and Methodist layman from the North, was already well known as a theistic evolutionist when the Vanderbilt trustees invited him in 1876 to join the faculty as a part-time lecturer, which required his spending only two months in the South each spring. All went smoothly until 1878, when Winchell published a little book titled *Adamites and Pre-Adamites,* in which he argued that humans had populated Earth long before the appearance of Adam in the Garden of Eden. Even more provocatively, he insisted that Adam had "descended from a black race, not the black races from Adam." When Winchell's views began circulating in the racially sensitive South, the trustees responded to criticism by abruptly abolishing his position. Publicly they justified their action on "purely economic grounds"; privately the Methodist bishop who headed the trustees informed Winchell that he had lost his job because of his opinion on "Adamites and Pre-Adamites," a position that did not require belief in evolution.[22]

The Methodists' Tennessee Conference applauded the trustees for courageously confronting the "arrogant and impertinent claims of . . . science, falsely so called," while at least one local newspaper sprang to Winchell's defense. The anticlerical *Popular Science Monthly,* published in the North, railed against the "stupid Southern Methodists" who used their "power to muzzle, repress, silence, and discredit the independent teachers of scientific truth." The editor, like many later commentators, overlooked the fact that evolution had had little to do with Winchell's brief stay at Vanderbilt, where by the 1890s "Christianized versions of Darwinian evolution" were again welcome on campus.[23]

Racism may have motivated some of Winchell's critics, but racial concerns rarely manifested themselves in the antievolutionary literature. One southern historian, writing about the evolution controversies of the 1920s, has noted that although objections to evolution "were usually religious in nature, the frequent references to the theory's heretical implications regarding man's ancestry suggested that much of the anti-evolution sentiment may have been related to traditional concepts of race held by a majority of the citizens of the state." However, the only evidence he cites in support of this claim is W. J. Cash's personal recollection that "one of the most stressed notions which went around was that evolution made a Negro as good as a white man—that is, threatened White Supremacy." This may have been true, but it is difficult to document from the publications of the time.[24]

Evolution in Southern Colleges and Universities

With the exception of Woodrow, no southern professor before World War I seems to have lost a job over Darwinism. And during the prewar period evolution frequently appeared in the classrooms of both state and sectarian colleges. Henry Clay White's experience at the University of Georgia illustrates the extent to which evolutionists could survive, at times even thrive, in the intellectual atmosphere of the New South. A professor of chemistry and geology, White joined the faculty in Athens in 1872. In 1875 he first cautiously endorsed evolution, and within a few more years he was freely teaching it to his students. Although Darwinism remained "heavily under fire from all sides," as one of his former students put it, White continued to expose his classes to it, and in 1887, without incident, he publicly declared himself to be an evolutionist. That same year a senior orator at the university delivered a positive address on evolution—stripped of any mention of its religious implications because of the school's ban on "any references to politics or religion upon the college stage."[25]

During the late 1870s or early 1880s White (or perhaps one of his colleagues in geology) commissioned a fresco depicting evolution for the ceiling of the geology lecture room. The artist created what a contemporary described as "a beautifully painted design, representing the evolution of life through all the geologic or zoologic ages." In 1909, when Darwin would have been a hundred years old, White planned a special birthday celebration in honor of the revered scientist. In deference to the concerns of the chancellor of the university, who feared criticism from evangelical antievolutionists, White hosted the event in his own home. Three of White's colleagues at the university—a historian, a biologist, and a classicist—joined him in honoring Darwin with prepared papers, as did a noted Episcopal bishop. Despite such enthusiasm for evolution, White remained in the good graces of the university until his death in the 1920s.[26]

At Tulane University in New Orleans theistic evolutionists also flourished, especially in the medical school, where there were three or four known Darwinists in the mid-1880s. J. W. Caldwell, Tulane's professor of geology and mineralogy, had resigned his position at Southwestern Presbyterian University (now Rhodes College) in Tennessee when the controlling synods in the wake of the Woodrow affair declared, as one cynic put it, "that Evolution is inconsistent with Synodical natural science." At Tulane, Caldwell suggestively advertised in the school catalogue that he aimed "to consider organic life, as it is expressed in the fossils of the various strata, and to discover, if possible, the connexion between the successive fauna and flora."[27]

John B. Elliott, professor of the theory and practice of medicine at Tulane, used his platform as president of the New Orleans Academy of Sciences in the late 1880s to promote evolution as a "great natural law" and to defend it against the charge that it led to unbelief. The son of an Episcopal bishop, he could at times barely contain his fervor for this new revelation from science. "The effect of the theory of natural selection upon the human mind," he declared on one occasion, "has been vivifying in the extreme, so bold, so clear-cut, and so simple; accounting for so much that can be accounted for in no other way." Despite his praise of natural selection, Elliott, like virtually all other scientific evolutionists in the South—and the vast majority in the North—preferred non-Darwinist modes of evolutionary development.[28]

At the University of Mississippi George Little, who had earned a Ph.D. at the University of Berlin, openly taught evolution, even though at the time of his arrival in 1881 "the controversy between evolution and orthodoxy was at its height." Never one to conceal his preferences, he liked to boast that his library contained "the works of Darwin, Huxley, Tindall [*sic*] and Spencer, alongside of those of Dr. McCosh of Princeton and Hugh Miller of Scotland." When a visiting general prayed that God would save the students at the University of Mississippi from "the teachings of 'science falsely so called' " and a short time later pointedly sent Little an article on evolution, the professor returned the piece with the terse comment that "evolution is a workable hypothesis, like Newton's law of gravitation and Dalton's Atomic Theory." In his opinion, evolution was also theologically harmless. In support of that claim he pointed out that many of his former students and one of his own sons had gone on to become ministers of the Gospel.[29]

At other public universities around the South—at Alabama, North Carolina, South Carolina, Virginia, and presumably elsewhere—the story was much the same: theistic evolution could be taught with relative impunity.[30] This does not mean that southern universities flaunted the teaching of evolution or that university administrators never advised caution. We know, for example, that in 1909 the dean of the University of South Carolina warned a prospective lecturer that "though we have made progress toward . . . evolution since Dr. Woodrow's time . . . we have hardly reached the point where we could make the subject too prominent." The dean expressed confidence that thoughtful people would not object, but he worried that narrow-minded religionists, particularly those associated with denominational colleges, would criticize the university if the visiting scholar touted evolution.[31]

Many religiously affiliated colleges in the South were themselves teaching evolution by the last decades of the nineteenth century. As Woodrow and his friends loved to point out, this was true even of a number of Presbyterian schools, such as Davidson College in North Carolina, Hamden-Sydney College in Virginia, Southwestern Presbyterian University in Tennessee, and Central University (which later merged with Centre College) in Kentucky, all of which at least taught evolution in "a purely expository manner" or used evolutionary texts. In fact, by the mid-1880s it seemed "doubtful whether any college deserving the name in the United States, North or South, uses a text-book on geology or biological science whose author is not an evolutionist, and in which Evolution is not taught."[32]

The fortunes of evolution at Southwestern Presbyterian are particularly revealing. Caldwell, the professor of the natural sciences there in the early 1880s, had come to accept evolution as "God's ordinary mode of working" even before Woodrow and had begun teaching theistic evolution in a nondogmatic manner. Because his "teaching did not assume a prominent or offensive form," explained the chancellor, he had seen no reason to bring Caldwell's activities to the attention of the board of directors. When the Woodrow bomb burst on the southern Presbyterian church, Caldwell voluntarily submitted his resignation and ran for cover to Tulane. Ironically, many of the same church leaders who opposed Woodrow's teaching in a seminary saw no reason for Caldwell to resign because, as one of Woodrow's supporters put it, "*he was a Professor in a* COLLEGE, [and] *there was no objection to him as a scientific man holding or teaching these views in a* COLLEGE!" This created an "almost incredible" situation, in which "our young men . . . may be taught by our scientific professors, even in our church schools, that Evolution is true; and then our professors in the theological schools must tell these same young men . . . 'Evolution is false.'"[33]

Evolutionists could also be found on Southern Baptist campuses. William Louis Poteat, a German-trained biologist on the faculty of Wake Forest College in North Carolina, helped to pioneer the teaching of scientific evolution in the South in the 1880s. Shortly after the turn of the century, Baylor University in Texas called one of Poteat's students, John Louis Kesler, to organize that school's biology department. He in turn recruited Lulu Pace, a theistic evolutionist, to join him in the department. Few other Baptist colleges at the time could afford to hire a professional biologist. On the eve of World War I Kesler could think of only two

Southern Baptist biologists, Poteat and Pace, who "would be thought of out of their own neighborhood when biology is mentioned." Both were widely known to be evolutionists, and neither had yet heard more than murmurings about their teaching Baptist students the theory of evolution.[34]

The Methodists' Wofford College in South Carolina and the Quakers' Guilford College in North Carolina were likewise exposing their students to evolution before the end of the century. During the 1890s two Guilford biologists, Joseph Moore and T. Gilbert Pearson, repeatedly and openly advocated organic evolution. Pearson, who went on to become a distinguished ornithologist, explained that life on Earth had begun with a single cell—a product of "the divinely appointed agencies of heat, gravity, chemical affinity, water, air and organic life"—which had evolved over "millenniums on millenniums" into the diversity of life seen today. At one point a committee of Guilford trustees investigated his orthodoxy, but Pearson presented such a compelling case for his Christian beliefs that he shortly thereafter received a 50 percent raise. He continued to teach evolution during his tenure at Guilford "and never heard any further complaint about my unorthodox views."[35]

Scientific Critics and the Charleston Circle

Some southern scientists, however, did criticize Darwin, especially during the first fifteen years or so after the publication of his *Origin of Species,* before a scientific consensus in favor of evolution developed. For example, the Louisville chemist and mineralogist J. Lawrence Smith, one of the region's most distinguished scientists, took a parting shot at Darwin in his address as the retiring president of the American Association for the Advancement of Science. But though he condemned Darwin for departing from "true science" in his "purely speculative studies," he professed to harbor no religious concerns. "If [evolution] be grounded on truth, it will survive all attempts to overthrow it," he declared; "if based on error, it will disappear, as many so-called scientific facts have done before." Similar sentiments also appeared from time to time in southern medical journals.[36]

Perhaps the most ardent scientific opponent of Darwinism in the South was John McCrady, a native of Charleston who spent the last years of his life as professor of biology and the relation of science and religion at the Episcopal University of the South in Tennessee. Before the Civil War

McCrady had studied with Agassiz at Harvard and had developed a special interest in embryology and radially symmetrical animals called Radiata. He had even, in 1860, published a preliminary paper on what he termed "the law of development by specialization," in which he hinted at the possibility of organic evolution and cautiously applauded Darwin's recent work. However, after the war, in which he suffered terribly, he began bitterly opposing "the current erroneous views of (so-called) 'Evolution,' " which he associated with Darwin. In 1873, at the invitation of Agassiz, McCrady joined the faculty of Harvard's Museum of Comparative Zoology, where he remained until the university president forced him to resign in 1877, ostensibly because of inadequate teaching and research but perhaps, as McCrady suspected, also because of his social and religious conservatism and his refusal to teach "Darwinism, Huxleyism and Spencerism." Returning to the South, he devoted the last years of his life to working on his law of development, which he grandiosely hoped would harmonize "the apparently antagonistic views of Agassiz & Darwin." But because of his reticence to publish his view on evolution, he exerted relatively little influence on southern thought.[37]

During the 1850s and 1860s McCrady had taught at the College of Charleston and had become an active participant in the circle of naturalists in the city who, especially before the war, earned a remarkable reputation for their scientific work. Within the circle, opinion regarding Darwinism was divided, with several prominent members refusing to be drawn into debate. The elderly John Bachman, a well-known Lutheran pastor and naturalist, condemned Lamarckian evolution but apparently remained silent about Darwin's theory. Francis S. Holmes, an Episcopalian who taught at the College of Charleston until 1869, also avoided the subject, though he sold Darwin's works in his bookstore. Lewis R. Gibbes, a fellow Episcopalian and longtime professor at the college, seems to have commented on evolution only once. Responding in 1891 to a request for his opinion on Joseph LeConte's efforts to reconcile evolution and Christianity, he jotted a note at the bottom of the letter saying that it was "best to avoid discordant discussion."[38]

At least two members of the Charleston circle looked favorably on Darwinism. Henry W. Ravenel, a noted mycologist, found some aspects of Darwin's theory unconvincing but nevertheless judged it "an attractive doctrine." Gabriel Manigault, a prodigious osteologist and longtime curator of the Charleston Museum, observed late in life that evolution was irrefutable. He thought it "strange" that so many contemporaries feared its "atheistic consequences." Besides McCrady, only J. H. Mellichamp, a botanist from Beaufort, South Carolina, with close ties to the circle, openly criticized Darwin. "The great Mr. Darwin, whom most worship," he informed a correspondent, is "the most inconclusive of all writers." Mellichamp wondered where Darwin would "be 50 years hence?—I wonder!—perhaps quite forgotten."[39]

As we have seen, scientific opposition to Darwinism did exist in the South before World War I, but for every McCrady and Mellichamp, there were perhaps several disciples of Darwin such as Silas McDowell and Moses Ashley Curtis, botanists in North Carolina and associates of Asa Gray, Darwin's chief American ally.[40] Outside of scientific circles there was considerably more anti-Darwinian sentiment, though some literate southerners appreciated Darwin for delivering "the final *coup de grace* to the untenable doctrine of the extreme rigidity of species and absolute invariability of types."[41] Even in the churches of the South one could sometimes hear voices urging toleration, if not acceptance. "Let the scientific men grapple with the hypothesis, they will deal with it according to its true merits," wrote a somewhat fearful Presbyterian; "and if it be true, it will ultimately take its place as an accepted scientific theory, in spite of the fulminations of the Vatican or the artillery of Protestant divines and metaphysicians." The Baptist New Testament scholar A. T. Robertson put it somewhat more colorfully in stating his openness to evolution with "'God' at the top." "I can stand it if the monkeys can," he would tell his students at the Southern Baptist Theological Seminary in Louisville.[42]

The Antievolution Crusade

The relative tranquillity evolutionists in the South enjoyed in the late nineteenth and early twentieth centuries declined rapidly in the years after World War I, when angry Fundamentalists, convinced that the teaching of human evolution was causing many of the nation's social ills, tried to dislodged evolutionists from their professorships and to ban the offending doctrine in public schools. Evolutionary biologists in Southern Baptist colleges suffered particularly harsh treatment during the witch hunts of the 1920s. Poteat at Wake Forest and C. W. Davis of Union University in Jackson, Tennessee, both survived efforts to oust them, but colleagues at Baylor, Mercer University in Georgia, and Furman University in South Carolina all fell victim to the Fundamentalist frenzy. Apparently no southern Presbyterians lost their professorships over evolution in the 1920s, perhaps because their denomination had learned a lesson from the embarrassing Woodrow affair, and the other major Protestant bodies generally remained aloof from the debates over evolution.[43]

Although state legislatures across the land debated the wisdom of antievolution laws, only three states—Tennessee, Mississippi, and Arkansas—actually outlawed the teaching of human evolution. In addition, Oklahoma prohibited the adoption of evolutionary textbooks, and Florida condemned the teaching of Darwinism as "improper and subversive." The fact that all these states were in the South enhanced the region's reputation for hostility to evolution and encouraged even some Fundamentalists to believe that the South was "the last stronghold of orthodoxy on the North American continent." The hoopla surrounding the Scopes trial in 1925 seemed dramatically to confirm the South's distinctiveness.[44]

Writers who describe the South as a bastion of antievolutionism typically neglect to mention that most southern legislatures refused to outlaw the teaching of evolution in the 1920s. This was true in Alabama, Georgia, North Carolina, South Carolina, Kentucky, Louisiana, and Texas, as well as in Florida and Oklahoma, where legislators declined to make the teaching of evolution a crime. In Virginia antievolutionists failed even to find a sponsor for their bill. Newspapers throughout the South, from the *Louisville Courier-Journal* and the *Atlanta Constitution* to the *Richmond News-Leader* and the *Raleigh News and Observer,* played a decisive role in helping to turn back the Fundamentalist onslaught. After the Alabama legislature defeated an antievolution bill in 1923, the editor of the *Montgomery Advocate* boasted that "every daily newspaper in the state, city daily and small daily also, opposed the bill, while the right to teach evolution has never been an issue in any college in Alabama—that is, not openly."[45]

A generally overlooked factor that contributed to the outburst of militant antievolutionism in the South was the growing popularity of Darwinism among the educated classes in the region. "Practically all biological teachers in the high schools and colleges (including Baptist colleges) believe in some form of biological development," wrote a Baptist educator from Kentucky in 1921. He estimated that "nine out of every ten of our Baptist preachers who are full college and seminary graduates believe in some form of biological development" and that "practically all of our grammar school, high school, and college graduates either believe as true or at least accept as plausible some sort of theory of biological development." These estimates may have been on the high side, but they would not have surprised J. Frank Norris, a fiery Fundamentalist preacher from

Fort Worth. "Contrary to what most people here in the South think," he informed William Jennings Bryan in 1923, "evolution has already made tremendous gains in our schools." For confirming evidence, he needed to look no farther than nearby Waco, where Baptist biologists at Baylor had been teaching organic evolution for years and where one sociologist had recently published a description of primitive man as "a squat, ugly, somewhat stooped, powerful being, half human and half animal, who sought refuge from the wild beasts first in the trees and later in caves." Exposed and vilified by Norris, the social scientist submitted his resignation. But the fact that a Southern Baptist professor had felt free openly to advocate human evolution tells us much about the degree to which Darwinism had penetrated the "mind" of South.[46]

4

THE SCOPES TRIAL: HISTORY AND LEGEND

Until the trial of O. J. Simpson captured worldwide attention in 1995, the Scopes trial, pitting William Jennings Bryan against Clarence Darrow, held title to the best-known legal contest in American history. At least that is what the authoritative *Encyclopedia of Southern Culture* assures us. Certainly the 1925 trial in Dayton, Tennessee, ranks among the most familiar events in the history of American science, rivaled perhaps by Benjamin Franklin's kite experiment and the making of the atomic bomb. Authors of textbooks who ignore the coming of Darwinism to America often devote a paragraph or two to the Scopes trial, typically to illustrate increasing tension between urban and rural values and to show how Bryan's bumbling performance discredited efforts to outlaw the teaching of evolution. The *National Standards for United States History: Exploring the American Experience* endorses this interpretation. The popular press takes the truth of this view for granted.[1]

Despite a shelf of scholarly studies on Fundamentalism, antievolutionism, and Bryanism, the Scopes trial remains a grotesquely misunderstood event—largely the result, I think, of its ability to serve so many competing interests. A glance at the secondary literature quickly dispels the sense of consensus one gets from looking at American history textbooks. The church historian Martin Marty dismisses the trial as an "irrelevancy," "a good show," while George M. Marsden assigns it a central role in discrediting Fundamentalism, and Robert D. Linder describes it as "a watershed in American religious history." Two recent books on creationism in America further illustrate the continuing historiographical disagreement about even the most basic issues. Raymond A. Eve and

Francis B. Harrold portray the Scopes trial as the climax of the antievolution movement, while George E. Webb, echoing Ferenc M. Szasz, uses it to mark "the beginning of the most active period of the controversy."[2]

The cultural context of the trial remains equally in dispute. Many historians persist in associating Fundamentalism and antievolutionism with a distinctively rural mindset, ignoring such colleagues as Szasz and Ernest R. Sandeen, who have stressed the northern, urban roots of Fundamentalism. Such leading antievolutionists as William Bell Riley, John Roach Straton, and Harry Rimmer hailed, respectively, from Minneapolis, New York City, and Los Angeles. In a unique empirical analysis of voting behavior in Arkansas, the only state in the 1920s to hold a public referendum on the evolution question, Virginia Gray a quarter-century ago found "that areas with a high concentration of rural, backward people were *not* the significant backbone of the anti-evolution movement." However, her discordant findings have been largely ignored.[3] And, as we saw in Chapter 3, historians who almost instinctively associate antievolutionism with the South almost never mention that the majority of southern legislatures rejected antievolution laws.

In this chapter I take a fresh look at the Scopes trial, focusing specifically on its representation (I am tempted to say misrepresentation) in popular and scholarly works. I would like to test some of the most widely held claims against the available evidence. Did the Scopes trial mark the end of the antievolution crusade? If not, how did this view come to enjoy such immense popularity? Did Bryan's testimony at Dayton destroy his credibility and the plausibility of special creation? How did contemporaries—Fundamentalists and modernists alike—interpret the celebrated event? Finally, I would like to explore the creation of competing legends about the trial and the interests that they served.

But, first, we need to review the facts of the case. In January 1925, by a vote of seventy-one to five, the Tennessee House of Representatives passed a bill making it unlawful for state-supported schools "to teach any theory that denies the story of the Divine Creation of man as taught in the Bible, and to teach instead that man has descended from a lower order of animals." The bill, proposed by the farmer-legislator John Washington Butler of Lafayette made violation of the law a misdemeanor carrying a fine of $100 to $500. Six weeks later the Senate approved the bill by a four-to-one majority, and shortly thereafter Governor Austin Peay signed the bill into law. Oklahoma legislators had already prohibited the adoption of textbooks that promoted evolution and Florida lawmakers had condemned the teaching of Darwinism as "improper and subversive," but Tennessee led the way in making the teaching of human evolution a punishable crime.[4]

In early May John Thomas Scopes, a young high school science teacher and coach in Dayton, agreed to test the constitutionality of the law. The fledgling American Civil Liberties Union (ACLU) in New York City, which had been soliciting volunteers for a test case, provided legal counsel, including the famed Chicago attorney and agnostic Clarence Darrow. William Jennings Bryan, thrice-defeated Democratic candidate for the presidency of the United State and leader of the Fundamentalists' efforts to ban the teaching of human evolution, joined the prosecution team.[5]

The trial itself began on Friday, July 10, and lasted eight days. The most dramatic moment came on the seventh day, when Bryan submitted to a lengthy examination by Darrow in which he affirmed his belief in the story of Jonah being swallowed by a "big fish," in Joshua's commanding the sun (or more probably Earth) to stand still, in Noah's flood, and in the Genesis story of creation. The outcome was never really in doubt, because even Scopes's defense attorneys conceded that he had taught the forbidden theory. In fact, they wanted a conviction in order to test the constitutionality of the law, believing that it violated not only the constitution of Tennessee but the Fourteenth Amendment of the United States Constitution, barring the states from making or enforcing "any law which shall abridge the privileges or immunities of citizens of the United States." The jury, after deliberating just nine minutes, found Scopes guilty, whereupon the judge imposed a fine of $100. Five days later Bryan died in his sleep. In January 1927 the Tennessee Supreme Court upheld the constitutionality of the Butler Act but overturned the conviction of Scopes on a technicality (the jury, not the judge, should have imposed the fine), thus depriving the defense lawyers of an opportunity for further appeal.[6]

The trial put the town of Dayton—and the state of Tennessee—on the international map. According to the *Nashville Tennessean*, "In Constantinople and far Japan, in Paris and London and Budapest, here and there and everywhere, at home and abroad, in pagan and in Christian lands, where controversialists gather, or men discuss their faiths, to speak the name of Dayton is to drop a bomb, to hurl a hand grenade, to blow the air of peace to flinders." The North Carolina sociologist Howard W. Odum estimated that the trial was discussed by "some 2310 daily newspapers in this country, some 13,267 weeklies, about 3613 monthlies, no less than 392 quarterlies, with perhaps another five hundred including bi-monthlies and semi-monthlies, tri-weeklies and odd types." His search had turned up "no periodical of any sort, agricultural or trade as well, which has ignored the subject." "For the first time in the history of American jurisprudence," noted the *Chicago Daily Tribune,* "radio went into a court of law and broadcast to the earphones and loudspeakers of all radio set owners who cared to listen to the entire proceedings of a criminal trial." It was, said some, "the world's most famous court trial."[7]

Bryan's Views

It is common knowledge that Bryan did not fare well in press coverage of the trial, especially in accounts of his verbal duel with Darrow. Two examples should suffice. "It has long been known to many that he was only a voice calling from a poorly furnished brain-room," observed an unsympathetic editorialist in the *New York Times.* "But how almost absolutely unfurnished it was the public didn't know till he was forced to make an inventory." The *Nashville Tennessean* quoted a defense attorney as saying that "after his examination by Clarence Darrow yesterday no intelligent person in American can believe that Mr. Bryan is an authority on science, education or the Bible."[8]

Historians and other critics of Bryan's performance have generally agreed that his clumsy defense of the Bible severely undermined his credibility, but they have been of divided mind in explaining what went wrong. Some have maintained that Bryan foolishly defended a recent creation in six twenty-four-hour days in the year 4004 B.C.; others have accused him of caving in under pressure from Darrow and conceding both the antiquity of Earth and the figurative meaning of the Genesis "days." Stephen Jay Gould, attempting to correct the erroneous views of others about the Scopes trial, calls Bryan's concession about the age of Earth "the most famous moment of the exchange." Bryan, he assures us, "did not seem to appreciate that local fundamentalists would regard [his acceptance of an ancient Earth] as a betrayal, and the surrounding world as a fatal inconsistency." However, Gould neglects to provide evidence that Fundamentalists regarded Bryan's testimony as a betrayal. A few confused writers, in their eagerness to show Bryan's muddle-headedness, have charged him with simultaneously defending an old and a young Earth.[9]

A simple examination of the trial record—which has been available in published form since 1925—readily clarifies Bryan's beliefs. Because the testimony has so often been misconstrued, it is worth reading in detail:

Q: Mr. Bryan, could you tell me how old the earth is?
A: No, sir, I couldn't.
Q: Could you come anywhere near it?
A: I wouldn't attempt to. I could possibly come as near as the scientists do, but I had rather be more accurate before I give a guess.

. . .

Q: Would you say that the earth was only 4,000 [*sic*] years old?
A: Oh, no; I think it is much older than that.

. . .

Q: Do you think the earth was made in six days?
A: Not six days of twenty-four hours.

. . .

Q: You think those were not literal days?
A: I do not think they were twenty-four-hour days.
Q: What do you think about it?
A: That is my opinion—I do not know that my opinion is better on that subject than those who think it does.
Q: You do not think that?
A: No. But I think it would be just as easy for the kind of God we believe in to make the earth in six days as in six years or in 6,000,000 years or in 600,000,000 years. I do not think it important whether we believe one or the other.
Q: Do you think those were literal days?
A: My impression is they were periods, but I would not attempt to argue as against anybody who wanted to believe in literal days.
Q: Have you any idea of the length of the periods?
A: No; I don't.

. . .

Q: The creation might have been going on for a very long time?
A: It might have continued for millions of years.[10]

It is clear from this testimony that Bryan not only rejected the notion of a 6,000-year-old Earth but freely interpreted the days of Genesis as vast periods of time. Such beliefs may have struck Darrow (and a host of historians) as being inconsistent with hard-core Fundamentalism, but there is little evidence that Fundamentalists themselves expressed either shock or surprise. During the 1920s Fundamentalists divided over the correct interpretation of the Mosaic account of creation, but few insisted on a young Earth. Of those who left opinions on the subject, the majority, following the revered *Scofield Reference Bible,* believed that Genesis 1 described two creations, one "in the beginning," the other eons later, about 4004 B.C., when God placed Adam and Eve in the Garden of Eden. In the "gap" between these two events mentioned in the Bible, Earth had witnessed a series of catastrophes and recreations recorded in the fossil-bearing rocks. An influential minority, including Bryan, chose to accommodate the fossil evidence by reading the "days" of Genesis as vast geological ages. Only a relatively tiny group, mostly Seventh-day Adventists, insisted on the recent appearance of life on Earth in six days of twenty-four-hours each. Despite their differences, the Fundamentalists in all three hermeneutical camps regarded themselves as strict biblical literalists.[11]

Among Fundamentalists a commitment to literalism did not rule out interpreting Scripture. In response to Darrow's query as to whether he claimed "that everything in the Bible should be literally interpreted," Bryan replied: "I believe everything in the Bible should be accepted as it is given there; some of the Bible is given illustratively. For instance: 'Ye are the salt of the earth.' I would not insist that man was actually salt, or that he had flesh of salt, but it is used in the sense of salt as saving God's people." The Bible itself taught that "one day is with the Lord as a thousand years" (2 Peter 3:8), and even the most conservative believers happily indulged in speculation about the real-life identities of the dragon, the seven-headed beast, and other exotica described in the Book of Revelation. In advocating the day-age interpretation of Genesis, Bryan found himself in impeccable Fundamentalist company. George Frederick Wright, author of an essay on evolution in *The Fundamentals,* subscribed to the same view, as did William Bell Riley, head of the World's Christian Fundamentals Association, the organization that had sent Bryan to Dayton. Fundamentalists may have argued over the relative merits of the day-age and gap interpretations, but differences on this issue were not an indication of heterodoxy. In the late 1920s Riley engaged the antievolutionist Harry Rimmer in two friendly debates on this very issue; neither questioned the other's orthodoxy.[12]

Fundamentalists should hardly have been suprised by Bryan's espousal of the day-age theory, because he had been publicly advocating it for years. In 1923, in a published response to allegations that he believed the creation had taken place in six days of twenty-four hours each, he insisted that he had "never, in writing or in my speeches, said anything to justify such a charge." He went on to explain that "According to the interpretation placed upon it by orthodox Christians, the day mentioned in the account of creation was of indefinite duration. The only persons who talk about a twenty-four-hour day in this connection do so for the purpose of objecting to it; they build up a straw man to make the attack easier, as they do when they accuse orthodox Christians of denying the roundness of the earth, and the law of gravitation."[13]

Had Fundamentalists known his *private* views on evolution, they might well have felt uncomfortable. Shortly before the trial he confided to a Fundamentalist physician friend, Howard A. Kelly of Johns Hopkins University, that he had no objection to "evolution before man but for the fact that a concession as to the truth of evolution up to man furnishes our opponents with an argument which they are quick to use, namely, if evolution accounts for all the species up to man, does it not raise a presumption in behalf of evolution to include man?" Until biologists could actually demonstrate the evolution of one species into another, he thought it best to keep them on the defensive. In the meantime he regarded prehuman evolution as silly rather than offensive.[14]

Despite the frequently repeated claim that Bryan betrayed his supporters when he acknowledged the evidence for an old Earth and the desirability of interpreting "days" as "ages," the only criticism I have found came from the margins of Fundamentalism: from the Seventh-day Adventist geologist George McCready Price and his small band of followers, who rejected the fossil evidence for the antiquity of life on Earth by attributing virtually the entire fossil record to Noah's flood. And, as we shall see in Chapter 5, they did so at a time when most Fundamentalists attached no geological significance to the deluge. On the eve of the trial Bryan had invited Price, "the principal scientific authority of the Fundamentalists," to assist the prosecution at Dayton and had identified him during Darrow's grilling as one of two scientists (the other being the deceased Wright) in whom he had confidence.[15]

Although Price could not attend the trial because of a teaching engagement in England, he wished Bryan well. His only advice was for the lawyer to avoid arguments "about the scientific or unscientific character of the evolution theory" and to focus instead on showing "its utterly divisive and 'sectarian' character, and its essentially anti-Christian implications and tendencies." For the latter task, he could think of no one "more capable" than Bryan. When Bryan failed to defend Price's peculiar account of Earth history during the trial, the self-trained geologist turned bitterly against the statesman. Shortly after the trial Price told a reporter that Bryan "really didn't know a thing about the scientific aspect of the case. He didn't care a fig either. He was too busy in his life to study." Years later Price attributed Bryan's "stultification and defeat" at the Scopes trial, "in which poor Bryan . . . was so piteously overwhelmed and defeated by the battery of evolutionary scientists under the clever leadership of the atheist Clarence Darrow," to his attachment to the day-age theory. Eager to escape blame for the public-relations disaster at Dayton, which he regarded as "a turning point in the intellectual and religious history of mankind," Price recast his role, falsely claiming that before the trial

> I wrote him at considerable length, pointing out how flimsy and easily refuted were the evolutionary arguments, when the Flood is taken as the cause of the fossils. And I urged him to adopt this policy or strategy, and *to be sure to put the opponents or the evolutionists on the defensive,* and make them prove their case, as he and his side clearly had a right to do. If he had done this, the history of the trial would certainly have been different.

Writing privately to an Adventist friend, Price attributed the fiasco at Dayton to the machinations of Satan himself: "The whole affair was undoubtedly planned long ahead by the devil; it was bungled by Bryan in almost every possible way; and the general results were a triumph for the devil's cause such as the world has seldom seen before or since."[16]

Other Fundamentalists expressed disappointment and anger over Darrow's shabby treatment of Bryan during the trial, but only Price and his fellow flood geologists faulted Bryan himself for straying from their ultraliteral reading of Genesis. Most Southern Baptists, reports James J. Thompson, Jr., saw the trial as a morality play in which "goodness had vanquished evil." Writing at the close of the trial, J. Frank Norris, the Baptist pastor from Fort Worth, likened Bryan to religious heroes of the past: "It is Moses challenging Pharaoh; it is Elijah arraigning Ahab; it is Paul defying Nero; it is Martin Luther hurling his thesis [*sic*] at Pope Leo X. It is the greatest battle of the centuries." William Bell Riley, who had sent Bryan to Dayton to represent organized Fundamentalism, scornfully denounced the blood-sucking journalists who covered the trial and Darrow's "unfair methods" in examining Bryan, but he expressed only pleasure at Bryan's "signal conquest" for Fundamentalism: "He not only won his case in the judgment of the Judge, in the judgment of the Jurors, in the judgment of the Tennessee populace attending; he won it in the judgment of an intelligent world." This is not the language of betrayal and disappointment.[17]

Fundamentalists were not the only ones upset by Darrow's behavior. The editor of the *Chattanooga News,* who had consistently opposed the Butler Bill, faulted Darrow for insulting the judge and denounced Darrow's cross-examination of Bryan as "a thing of immense cruelty." He surmised that it would damage attempts to repeal the law by creating a backlash of sympathy for Bryan. Darrow also received public reprimands for gratuitously protesting the opening of court with prayer, for not quietly testing the constitutionality of the Butler Act in chancery court, as was the custom in Tennessee, and for undermining local democracy in favor of national norms. As a writer in the *New Republic* pointed out, the ACLU had "as much reason to affirm the liberty of the people of Tennessee to control the curriculum of their public schools as it had to affirm the liberty of Mr. Scopes to teach the doctrine of evolution in those schools." In his opinion, "The right of Tennessee to pass a foolish law is quite as much in need of defense as the right of a soap-box orator to utter a foolish or a dangerous speech." As the journalist Walter Lippmann saw it, Bryan and the Tennessee legislature were simply applying the Jeffersonian principle that "it was sinful and tyrannical to compel a man to furnish contributions of money for the propagation of opinions which he disbelieves. . . . Jefferson had insisted that the people should not have to pay for the teaching of Anglicanism. Mr. Bryan asked why they should be made to pay for the teaching of agnosticism."[18]

By the end of the trial Darrow was trying the patience of even his supporters. Darrow and his colleagues were "rather getting on the nerves of sober-minded people of this country," observed the *Boston Post.* Especially grating was their "assuming an air of lordly superiority over lawyers and judges or a state of whose jurisprudence they know next to nothing." Non-Fundamentalist criticism of Darrow grew so intense after the trial that the ACLU tried to dump him from the team of defense lawyers appealing Scopes's conviction to the Tennessee Supreme Court. The Chicago attorney, however, refused to step down. John R. Neal, chief counsel for Scopes, complained that the defense had "received little but criticism from a number of liberal Christian magazines," which generally opposed Tennessee's antievolution law. In the end, concluded the liberal *Christian Century,* Darrow had proved as great "an embarrassment to the cause he insisted on championing" as had Bryan.[19]

The Beginning or the End?

Looking back at the Scopes trial, numerous observers, convinced that Darrow had thoroughly discredited Bryan, have described it as the effective end of the antievolution crusade, perhaps of the Fundamentalist movement itself. The trial "brought feeling to a head and provided a dramatic purgation and resolution," wrote the historian Richard Hofstadter. "After the trial was over, it was easier to see that the antievolution crusade was being contained and that the fears of the intellectuals had been excessive." Contemporaries, however, tended to see events differently. In examining coverage of the trial in five geographically scattered newspapers and over a dozen national magazines, I discovered not a single declaration of victory by the opponents of antievolutionism, in the sense of their claiming that the crusade was nearing an end.[20]

After Scopes's conviction in court, no one, of course, could claim that the defense had won a legal victory. But some opponents of Fundamentalism did declare other sorts of victories. On the basis of exposing the weaknesses of the opposition, Dudley Field Malone claimed a "victorious defeat" for the defense. Philip Kinsley, writing for the *Chicago Tribune,* suggested that the Scopes defense had won an "intangible victory" by exposing so many young people in Tennessee to the outlawed theory of evolution. But as Paul Waggoner has shown (in a seldom-cited but important study), the notion of a Fundamentalist defeat at Dayton did not take hold until the early 1930s, when the journalist Frederick Lewis Allen published his best-selling *Only Yesterday: An Informal History of the Nineteen-Twenties* (1931). According to Allen, "Theoretically, Fundamentalism had won, for the law stood. Yet really Fundamentalism had lost. Legislators might go on passing anti-evolution laws, and in the hinterlands the pious might still keep their religion locked in a science-proof compartment of their minds; but civilized opinion everywhere had regarded the Dayton trial with amazement and amusement, and the slow drift away from Fundamentalism certainly continued." After Allen, the legend that Bryan and the Fundamentalists had lost at Dayton became a truism.[21]

Nothing has done more to popularize this legend than *Inherit the Wind.* This classic film, released in 1960, is based on a 1955 play written by Jerome Lawrence and Robert E. Lee. To the strains of "Give Me That Old-Time Religion," the movie shows the arrest of a young biology teacher, Bertram Cates (John Thomas Scopes), who violates a law banning the teaching of human evolution. The subsequent Monkey Trial, held in the small town of Hillsboro (Dayton), "the buckle on the Bible Belt," pits the famous Chicago lawyer and agnostic Henry Drummond (Clarence Darrow) against the Fundamentalist politician Matthew Henry Brady (William Jennings Bryan). Orchestrating the show is the caustic E. K. Hornbeck (H. L. Mencken), a Baltimore journalist, whose newspaper pays for Cates's defense. The acting is superb; the moral—the danger of right-wing witch hunts—is timely.[22]

But is the movie good history? As Scopes himself noted in his autobiography, *Center of the Storm* (1967), the filmmakers took a number of liberties with "the facts of the real trial." For example, to heighten the drama and to provide a little romance for what Scopes called "the balcony set," the movie shows Cates sitting in jail and courting the preacher's daughter, who is torn between faith and love. Scopes himself spent no time behind bars and was never formally arrested; a local pharmacist, who chaired the school board, simply informed a Chattanooga newspaper that Scopes had been "arrested" (a term Scopes himself always put in quotation marks). Town fathers did not walk into his biology classroom and catch him red-handed teaching the evolution of humans from "a lower order of animals," as the film suggests. Scopes, who had recently received an undergraduate law degree from the University of Kentucky, coached football and basketball and taught algebra, physics, and chemistry—but not biology—at the Dayton high school. He volunteered to test the anti-evolution law on the off chance that he had mentioned evolution while filling in for the regular biology teacher during a review session. His inability to recall having ever covered evolution, despite the (perhaps unreliable) testimony of several students that he had, contributed to the defense's decision not to put him on the witness stand during the trial.[23]

In spite of the artistic license taken by the playwrights, Scopes believed that the film accurately "captured the emotions in the battle of words between Bryan and Darrow." Perhaps so, but it grossly caricatured the stated opinions of the protagonists. It especially distorted Bryan's (that is, Brady's) views on Genesis and geology, by having him appeal to "Bishop Ussher" as authority for believing that the creation took place at 9:00 A.M. on October 23, 4004 B.C. The playrights highlighted Brady's subsequent admission that the days of creation could have been longer than twenty-four hours with the following staging instructions: "Drummond's got him. And he knows it! *This is the turning point.* From here on, the tempo mounts, Drummond is now fully in the driver's seat." I have no quarrel with the playwrights' invention of dialogue to heighten the drama of their fictional encounter. I do, however, question the judgment of the historians who drafted the *National Standards for United States History,* which recommends the movie as an aid to understanding Bryan's mind and "fundamentalist thinking" generally. This strikes me as being a little like recommending *Gone with the Wind* as a historically reliable account of the Civil War.[24]

Some theological modernists may have viewed the trial in Dayton as the turning point in their war against Fundamentalism, but Fundamentalists themselves emerged from the trial flushed with a sense of victory. Before his death Bryan vowed to "strike while the iron is hot." According to newspaper reports, he intended to "carry his holy war against science into the legislatures of seven more states within the next two years," guided by the "star of Dayton." Shortly after Bryan's death, John Roach Straton, who hoped to inherit his mantle, promised that "within a year there would be a struggle in nearly every State." While touring the South in late September 1925, he detected signs of "a great Revival of Religion" forming in the wake of the Scopes trial. Edward Young Clarke, erstwhile Imperial Wizard of the Ku Klux Klan and founder of the Supreme Kingdom in 1926, announced that within "two years, from Maine to California and from the Great Lakes to the Gulf, there will be lighted in this country countless bonfires, devouring those damnable and detestable books on evolution."[25]

Leading anti-Fundamentalists saw it the same way. In his last dispatch from Dayton to the *Baltimore Evening Sun,* H. L. Mencken declared that Genesis had emerged "completely triumphant" from the battle. "There are other States," he advised, "that had better look to their arsenals before the Hun is at their gates." At the conclusion of the trial the defense attorney Dudley Field Malone forecast a renewed effort by Bryan to stifle the teaching of evolution, while his colleague John R. Neal observed that the movement to pass antievolution laws had "back of it strong organizations in every State." The New York Unitarian Charles Francis Potter, who had traveled to Dayton to assist Scopes, warned that the "trouble over evolution" had "only just started." The Fundamentalists' "success at Dayton" was so complete, *The Nation* predicted that 1926 would witness "a flood of bills introduced into the legislature of every State in the Union."[26]

Events following the trial proved the worriers right. When Bryan died just five days after the close of the trial, depriving "the fundamentalist menace" of its undisputed leader, some evolutionists assumed that the campaign against evolution would quickly go into decline. But rather than slowing down, it picked up momentum. As one British magazine reported, the Fundamentalists' "almost fanatical faith has been set ablaze by the dramatic death of Mr. Bryan." In North Carolina, for example, Bryan's death "became the rallying point for the anti-evolutionists," notes the historian Willard B. Gatewood, Jr. "Martyred in behalf of a noble cause, the Great Commoner from his grave breathed new life into the anti-evolutionist crusade . . . by enhancing the cohesiveness of the movement and by strengthening its determination to outlaw evolution as a means of avenging his death."[27]

In 1926, a year in which relatively few state legislatures met, Mississippi joined Tennessee in outlawing the teaching of human evolution. This prompted Virginius Dabney, of the *Richmond Times-Dispatch,* to predict another wave of antievolution legislation. The Fundamentalists, "their appetites whetted by victories in Tennessee and Mississippi, are thirsting for more blood," he wrote. "They seem determined to save the Republic from the devil." The chancellor of Washington University in St. Louis predicted ominously that "fifteen or twenty States" would pass antievolution laws in 1927. That hyperbolic threat never materialized, but the antievolution crusade did peak that year, when over a dozen states—from Maine to California—considered restrictive legislation. In 1928 Arkansas became the third (and last) state to ban the teaching of human evolution. By that time the movement was beginning to wind down. Maynard Shipley, of the Science League of America, who closely monitored antievolution initiatives, reported at the close of the decade that thirty-seven "anti-evolution bills, resolutions or riders" had been introduced in twenty different state legislatures during the years from 1921 through 1929. Of these, twenty-three (62 percent) dated from after 1925. In the face of these statistics, it seems just plain wrong to claim that the events in Dayton in July 1925 brought the antievolution crusade to a close—or even slowed it down.[28]

By 1929 only two states, Texas and Oklahoma, were still debating the merits of antievolution legislation. Given this rapid decline in interest, Shipley feared that the general public would conclude that "the fight is over, and science has won." But, he warned,

> Nothing could be further from the truth. The Fundamentalists have merely changed their tactics. As one of their leaders has worded it,

"We were too precipitate; we must go directly to the people themselves and not depend on the legislators." Primarily, they are concentrating today on the emasculation of textbooks, the "purging" of libraries, and above all the continued hounding of teachers.

Indeed, following the Scopes trial several textbook publishers quickly modified their treatment of evolution. They deleted potentially offensive statements or substituted the euphemism "development" for "evolution." In the absence of legislative action, the Texas state textbook commission banned the use of books that taught evolution, as did the Atlanta, Georgia, school board. Thus the Fundamentalists emerged victorious on both the legal and the pedagogical fronts.[29]

Although the Fundamentalists never gave up their goal of extirpating evolution from the churches and schools of America, they had by the end of the 1920s abandoned their efforts to outlaw the teaching of human evolution by legislative fiat. Historians have identified a host of factors contributing to this shift: from "sheer boredom" to the Great Depression. The historian Edward J. Larson has suggested perhaps the most compelling reason: "the string of legislative defeats suffered in the late 1920s," which convinced antievolutionists that they had little to gain from continuing their assault on statehouses. Instead, they turned their attention to local schoolhouses.[30]

Fundamentalist Legends

As one might expect, creationists have constructed their own legends about the Scopes trial and its consequences. A few creationist writers have followed the disappointed George McCready Price in portraying the trial as "evolution's great triumph." According to Henry M. Morris, the most influential of Price's latter-day disciples, Bryan, ignoring Price's advice, "foolishly" agreed to undergo Darrow's questioning. When Bryan endorsed the day-age theory of Genesis, "the creationist revival of the 1920s was all but stopped dead in its tracks." Jerry Falwell, perhaps the best known Fundamentalist in late-twentieth-century America, has adopted a similar interpretation, arguing that Darrow's examination of Bryan "proved disastrous for Bryan and Fundamentalism." Bryan, he writes, "lost the respect of Fundamentalists when he subscribed to the idea of periods of time for creation rather than twenty-four-hour days."[31]

As we have seen, most Fundamentalists in the 1920s tended to view Bryan's performance at Dayton in a much more positive light, and heroic images of Bryan have continued down to the present. One popular story tells how he stumped the scientific experts at the trial. As early as 1929 the Fundamentalist evangelist and self-styled research scientist Harry Rimmer ridiculed one of the defense's authorities on evolution, H. H. Newman of the University of Chicago, for claiming to have found evidence in Bryan's home state of Nebraska of a million-year-old man, dubbed Nebraska Man or *Hesperopithecus.* "This man and his age constituted scientific argument that was incontestable, and Mr. Bryan had no reply, except to say he did not believe that the existence of such a man had ever been proved. At which the experts laughed him to scorn, for the proof was incontestable!" However, it turned out after the trial, reported Rimmer, that the so-called proof, a single fossil tooth, had come from an extinct pig. Over half a century later the story was still making the rounds among creationists.[32]

Like most legends, however, the story has only a tenuous connection to historical events. First, Newman never testified in court, and the written statement he submitted said nothing about Nebraska Man. The legend, as John Wolf and James S. Mellett have argued, apparently originated in an exchange between the paleontologist Henry Fairfield Osborn and Bryan before the trial, not between Newman and Bryan during the trial. After Osborn announced the discovery of Nebraska Man in 1922, he taunted Bryan by suggesting that the ancient animal should have been named *Bryopithecus,* "after the most distinguished Primate which the State of Nebraska has thus far produced." Bryan, in turn, chalked Nebraska Man up to Osborn's "inflamed imagination." Osborn apparently discovered that he had confused a primate with a pig on the eve of the trial, which may have contributed to his decision not to go to Dayton.[33]

Some creationists in recent decades have attempted to refurbish Bryan's tarnished public image. In an influential essay on the trial, published in 1969, the creationist biologist Bolton Davidheiser challenged the received opinion "that the anti-evolution forces won the trial but lost the case." He portrayed Bryan as "the hero of the Scopes trial," who, though "repeatedly insulted and humiliated by Mr. Darrow, . . . did not complain or retaliate." And rather than displaying his ignorance, "he demonstrated that he had a better basic understanding of evolution and of the Bible than his detractors had."[34]

It is hardly surprising to find present-day creationists restoring Bryan to his rightful place in the Fundamentalist pantheon, but it is something of a shock to discover them preparing a pedestal for Bryan's archenemy Clarence Darrow. Yet creationists fighting for equal time in the schools have welcomed Darrow as a fellow traveler for arguing "that teaching only one theory of origins is sheer bigotry." The creationists' legal guru, Wendell R. Bird, has picked up on this theme, as has their most prominent evolution debater, Duane T. Gish, who claims that "Clarence Darrow thundered that it was bigotry to teach only one theory of origins." Actually the defense attorney Dudley Field Malone, not Darrow, made the most eloquent plea for teaching students both sides of the issue: "For God's sake let the children have their minds kept open—close no doors to their knowledge; shut no door from them. Make the distinction between theology and science. Let them have both. Let them both be taught. Let them both live." Now that evolutionists have gained the upper hand in the battle for curriculum time, such pleas generally come from the lips of creationists, not ACLU lawyers.[35]

Historical Lessons

If we strip away the many myths attached to the Scopes trial, what can we say about it as a historical event? Or, perhaps, what should we no longer say? Clearly the trial did not bring the antievolution movement to an end or even slow it down. The high point of the crusade did not come until two years later. Because Bryan's espousal of the day-age interpretation of Genesis 1 represented Fundamentalist orthodoxy at the time, he did not betray his cause while on the witness stand. He did not cave in under pressure from Darrow; he merely stated his long-held views. The trial did subject Bryan and Fundamentalism to considerable public ridicule, and it did contribute in some circles to a negative, antiscientific image of the South. It no doubt encouraged publishers to downplay evolution in their texts. But most of all it served as a protean source for legend-making: about Bryan and Darrow, about Fundamentalism and modernism, about science and religion. What the trial has come to represent is far more important historically than what the trial accomplished.

5

"SCIENCES OF SATANIC ORIGIN": ADVENTIST ATTITUDES TOWARD EVOLUTIONARY BIOLOGY AND GEOLOGY

> Again and again we shall be called to meet the influence of men who are studying sciences of satanic origin, through which Satan is working to make a nonentity of God and of Christ.
>
> Ellen G. White, *Testimony for the Church,* No. 37

> There ought to be no doubt whatever that the popular forms of geology and paleontology should be included as "sciences of satanic origin."
>
> George McCready Price, *Theories of Satanic Origin*

Seventh-day Adventism arose out of the ashes of the Millerite movement of the 1840s, when William Miller, a Baptist farmer-preacher from upstate New York, aroused the nation with his prediction that Christ would return to Earth in 1843 or 1844. Following the "Great Disappointment" of October 22, 1844, the date on which many Millerites had pinned their hopes, the movement splintered into several factions. One of these, led by the teenage visionary Ellen G. White and her husband, James, evolved in the early 1860s into the Seventh-day Adventist Church, distinguished by its belief in the imminent second coming of Christ, its observance of Saturday as the Sabbath, and its belief in the divine origin of Mrs. White's trance-like "visions."[1]

Ellen G. White and Mosaic Science

The birth of Seventh-day Adventism coincided with the opening of the nineteenth-century debate over evolution. In 1844, the year of the Great Disappointment, the Scottish author and publisher Robert Chambers anonymously published his *Vestiges of the Natural History of Creation,* the book that first brought the subject of evolution to the attention of the American reading public. Fifteen years later, on the eve of denominational organization, Charles Darwin brought out his monumental *Origin of Species.* By this time many American Christians, including evangelicals, had broken free from the notion of a 6,000-year-old Earth and a geologically significant deluge at the time of Noah and had reinterpreted the first chapters of Genesis to accommodate the growing geological evidence for an ancient Earth, far older than a few thousand years.[2] Seventh-day Adventists, however, staunchly defended the recent appearance of life on Earth and a universal flood that had deposited most of the fossils.

By far the most influential early Adventist was Ellen G. White, whose visions and testimonies molded the sect's thinking on matters ranging from diet to eschatology. Like her fellow believers, few of whom had been exposed to the influence of higher education, Mrs. White consistently subordinated science to Scripture. "The Bible is not to be tested by men's ideas of science," she wrote, "but science is to be brought to the test of this unerring standard." Because Moses had written his account of creation "under the guidance of the Spirit of God," any theory contradicting it was to be rejected out of hand. As far as she was concerned, Moses had left no doubt that the days of creation were six in number and of twenty-four hours duration and that God had not employed natural laws as the mode of creation.[3]

The editors of the *Review and Herald,* the group's official paper, shared Mrs. White's views on the relationship between science and religion. Early in 1859, months before the publication of the *Origin of Species,* they reprinted an excerpt from a non-Adventist source claiming that "while the Bible does not teach science, when it does refer to science it is always correct." This theme appeared frequently in early Adventist literature.[4]

One of the few warnings against an unreasoning dependence on the Bible in matters of science came from a member of the small educated minority in the church, a young physician named John Harvey Kellogg, who later won fame as the inventor of flaked cereals. A recent graduate of the Bellevue Hospital Medical College in New York City, Kellogg served as superintendent of the Adventists' sanitarium in Battle Creek, Michigan, and as professor of physics in their newly founded college. Writing in 1879 in a small volume entitled *Harmony of Science and the Bible,* Kellogg listed as one of the chief factors responsible for the recurring conflict between religion and science the habit among Christians of "holding the Bible as unimpeachable authority on all subjects, as the universal test of truth, and attaching all importance to a particular interpretation of its language." Although Kellogg apparently believed in a special creation, he expressed a willingness to recognize the legitimacy of science within its own sphere. "Science deals chiefly with one sort of truths, religion with another class of truths." If only this division were honored, he thought all conflict would cease.[5] In an article in the *Health Reformer,* which he edited, he called for "careful scrutiny of the arguments" in favor of evolution and an end to jeers and sneers. While adhering to the special creation of species in the beginning, he conceded that much of Darwin's theory was based on "indisputable" facts.[6]

Adventists typically placed their faith in the Bible rather than science because of a deep suspicion of human reason, and nothing seemed to confirm this suspicion better than the science of geology, which depended crucially on the assumption of uniform natural laws. Thus while the scientific and religious leaders of America were discussing the merits of biological evolution, Adventists were often preoccupied with the real or imagined fallacies of geology, which they saw as providing a foundation for organic evolution—both theories going "hand in hand to destroy faith in the word of God." Seldom did they miss an opportunity to point a scoffing finger at "the dreamy, incoherent utterances of geologists." Uriah Smith, editor of the *Review and Herald,* occasionally led the attack himself. Though he had never attended college, he had no fear of doing battle with the Goliaths of the scientific world. Who, he asked, had "ever proven or tried to prove" the validity of the principle of uniformity? "Nobody," he answered. "Usually it is either 'presumed that the reader will be convinced' of the matter, or certain results are 'supposed to have been effected by such causes as are operating at present.' "[7]

From time to time Smith opened the pages of the *Review and Herald* to other critics of geology. Their titles alone reveal their message: "The Blunders of Geologists," "The Uncertainty of Geological Science," and "False Theories of Geologists." Typical was the comment of George W. Amadon, the twenty-eight-year-old editor of the *Youth's Instructor,* a periodical for Adventist young people. "No class of scientific men are more hasty and rash in making assertions than some geologists," he wrote. "As a science it is not demonstrative, and its oracles are contradictory and clash with each other."[8]

Smith and his colleagues frequently reprinted what they considered to be devastating examples of the "extravagant pretensions" and the "absurdity" of geology. In one of these, Robert Patterson, a conservative Presbyterian minister who shared the Adventists' hope that Christ would soon return, remarked that to construct Earth's history from processes currently observable was like measuring "a youth of six feet high, and finding that he grew half an inch last year, [concluding] thence that he was a hundred and forty-four years old." In another, President Joseph F. Tuttle of Wabash College was said to have scored "a capital hit on that popular farce and prime minister of skepticism, geological guess-work," when he suggested that fossils—particularly human ones—found in geological formations much lower and earlier than usually assigned to humans had probably dropped to that level during earthquakes.[9]

Among the sizable group of Adventists to comment on geology, not one had any firsthand acquaintance with the science, and few gave any evidence that they had read more than popular accounts of what geologists did. Perhaps the best informed was Alonzo T. Jones, a self-taught ex-soldier who converted to Adventism while stationed at Fort Walla Walla, Washington, and became a minister. Jones took geology seriously enough to read Archibald Geikie's *Textbook of Geology,* one of the most authoritative works in the field, three times through. All this study, however, merely convinced him of the total unreliability of geology, a theme he developed at length in a series of lead articles for the *Review and Herald* in 1883. In them he accused geologists not only of beginning their reasoning with an assumption, but also of using circular arguments. The most blatant instances of the latter, he thought, were two statements by Geikie on dating. "One of these says that the *relative age of the rocks* is determined *by the fossils.* The other says that the *relative age of the fossils* is determined *by the rocks.*" "What is this but reasoning in a circle?" asked Jones. This example and others like it led him to conclude that "the only *certain* thing about [geological science], is its *uncertainty.*"[10]

Most Adventists refused to admit harboring any hostility toward what they liked to call "true science," that is, science based on "facts" and in agreement with the Bible. They directed their criticism solely at "science falsely so-called," hypothetical science in conflict with revelation. Scientific theories and hypotheses regarding the history of Earth were acceptable only under the severest restrictions. In formulating them, scientists were not to "assume any condition of the world, the existence of any agents, or the occurrence of any events, the reality of which they cannot demonstrate; and all their assumptions and reasonings must be consistent with all the facts, and all the laws of nature, which the question affects." It did not disturb Adventists that these stipulations also ruled out as unscientific all supernatural explanations of the creation of the world. They happily transferred the entire question of origins from the sphere of science to the realm of faith.[11]

In defending their extreme inductivist view of science, some Adventists revealed an anti-intellectual prejudice, not uncommon among Americans with little education. In 1872 the *Review and Herald* reprinted an address by the Presbyterian minister John Hall, in which he warmly thanked scientists for collecting so many useful facts, then denied them an exclusive right to interpret what they had discovered. "When they come to reason upon these facts," he said,

> they use just the same kind of mind that God has given me; and I endeavor to use my mind upon these facts aright, just as truly as they claim to use their minds upon the facts. Hence . . . I claim the right to reason upon them just as truly as they can claim it; and I do not think the less of myself if in many instances I draw conclusions from the facts that have thus become common property that are not the conclusions that they venture to draw!

Most Adventists could not have agreed more.[12]

Nevertheless, Adventist opposition to developmental theories, both organic and inorganic, focused not on the uncertain status of these ideas but on their apparent conflict with revelation. Because the Bible clearly stated that God had made the world in "six natural days," Adventists rebelled at the thought of sacrificing this divine truth "on the altar of geological speculation." Few American evangelicals spelled out the implications of such a sacrifice more sharply than David Nevins Lord, a millenarian Congregationalist from New York whose byline sometimes appeared in the *Review and Herald* and whose book *Geognosy; or, The Facts and Principles of Geology against Theories* served for a time as a geology text at the Adventist college in Battle Creek. Genesis and geology, he asserted, are mutually contradictory. If the geologists are correct, the Mosaic record is false and God is a liar—and "it is impossible that God should not have spoken the truth." The decision to accept or reject geology thus took on profound theological significance. "If founded on just grounds, [geology] disproves the inspiration, not only of the record in Genesis of the creation, but of the whole of the writings of Moses, and thence . . . of the whole of the Old and New Testaments, and divests Christianity itself of its title to be received as a divine institution."[13]

Compounding the difficulty of harmonizing any developmental view with the Bible were the statements of Mrs. White. Writing in 1864, she claimed that God had shown her in a vision the actual creation of the world. During this out-of-body experience she had seen "that the first week, in which God performed the work of creation in six days and rested on the seventh day, was just like every other week." For many Adventists the rejection of her testimony would have been tantamount to repudiating God's own word.[14]

Adventists especially feared anything that might weaken their arguments for observing Saturday rather than Sunday as their Sabbath. They frequently quoted the fourth commandment of the Decalogue found in Exodus 20 (but never the version in Deuteronomy 5), which mandated that the seventh day be kept holy, "because in six days the Lord made heaven and earth, the sea, and all that in them is, and rested the seventh day: wherefore the Lord blessed the sabbath day, and hallowed it." Theories of evolutionary development threatened to undermine this rationale. According to Ellen White, "the infidel supposition, that the events of the first week required seven vast, indefinite periods for their accomplishment, strikes directly at the foundation of the Sabbath of the fourth commandment." Her husband, James, president of the church, likewise warned that any deviation from the traditional view of creation would undermine the doctrine of the Sabbath along with the rest of the Bible. If the days of creation were assumed to be long, indefinite stretches of time, then "the period of man's toils and cares before a day of rest, is also immense, covering millions of years. And if the last day of the first week, the day on which Jehovah rested from his work, was another immense indefinite period, the weekly Sabbath of the Old and New Testaments, which was made for man and commanded in the moral law to be kept holy, is also an immense period of time." Such ideas, making the Bible seem absurd, obviously could not be tolerated.[15]

The only accommodation to an expanded time scale Adventists ever entertained was the possibility of allowing an extended period of time between an initial creation of inorganic matter "in the beginning," depicted in the first verse of Genesis 1, and a later six-day creation about 6,000 years ago. In the opinion of at least one Adventist, a midwestern minister named J. P. Henderson, this view did "no violence to a single statement in the Bible." Yet, despite its innocuousness, this compromise never gained much popularity among Adventists except for a brief period near the turn of the century, when both the *Review and Herald* and its sister magazine *Signs of the Times* came out in support of an old-Earth theory. But the prevailing attitude remained that earlier expressed by J. N. Andrews, a church leader, who insisted that there was "blank nothing" prior to the week of creation: "Even the materials which subsequently formed the worlds had no existence."[16]

Their strict adherence to a literal reading of Genesis kept most Adventists from adopting the interpretations, increasingly popular even among other evangelical Christians, that stretched the history of life on Earth to millions of years. Adventists also ran counter to intellectual trends in insisting on miracles as the sole mode of creation. By the second half of the nineteenth century many other Protestants were placing less emphasis on supernatural interventions in the natural order and more on God's general providence through the secondary laws of nature. This made it possible for them to explain evolution simply as God's way of creating the world through the mechanism of natural laws. Adventists, however, saw evolution as restricting, if not altogether abolishing, God's role in the work of creation. It "is the last and most plausible attempt of infidelity to vote the throne of the adorable Creator vacant," wrote one contributor to the *Review and Herald.* Another described evolution as "only an attempt to eject God, and to postpone him, and to put him clear out of reach."[17]

Because of the allegedly impious tendencies of evolution, Adventists commonly labeled it "atheistic" or "infidel," and its founders and supporters fared no better. The *Review and Herald,* for example, unapologetically published Thomas Carlyle's description of Darwin as an unintelligent atheist and reprinted a statement that "all the leading scientists who believe in evolution, without one exception the world over, are infidel." The fact that theistic evolution had won widespread acceptance in the Christian world—being "almost all-pervading in the orthodox and evangelical churches, schools, and colleges," reported one observer hyperbolically—carried no weight with Adventists. It merely provided additional evidence of the apostasy afflicting the nation's leading churches, explained W. H. Littlejohn, the president of Battle Creek College.[18]

Nontheological considerations played a secondary, but significant, role in the Adventist resistance to organic evolution. Human vanity rebelled at the prospect of relinquishing an honored position at the head of created beings only to be herded together "with four-footed beasts and creeping things," over which man had formerly had dominion. Darwinism, complained one unhappy critic, "tears the crown from our heads; it treats us as bastards and not sons, and reveals the degrading fact that man in his best estate—even Mr. Darwin—is but a civilized, dressed up, educated monkey, who has lost his tail."[19] For Adventists, who believed they had been created in the image of God himself, the demotion was indeed humiliating.

Those who rejected the evolutionary history of life necessarily had to provide an alternative explanation of the fossil record, and here Adventists invariably turned to the Noachian flood for virtually all solutions to their geological and paleontological problems. Encouragement to do so came from Ellen White, who wrote that if individuals would only recognize "the size of men, animals and trees before the flood, and . . . the great changes which took place in the earth," they would have no trouble accepting the "view that creation week was only seven literal days, and that the world is now only about six thousand years old." She believed that some recent geological findings suggestive of a great flood were providential, designed by God to "establish the faith of men in inspired history." Following her lead, the editors of the *Review and Herald* widely publicized any new discoveries that might conceivably corroborate the occurrence of Noah's flood.[20]

Adventists adopted a similar approach in defending Mrs. White's explanation of the great variety of life: amalgamation between species, including humans and animals. According to her, "every species of animal which God had created were [*sic*] preserved in the ark. The confused species which God did not create, which were the result of amalgamation, were destroyed by the flood. Since the flood there has been amalgamation of man and beast, as may be seen in the almost endless varieties of species of animals, and in certain of men." This statement—especially the allusion to crossbreeding between humans and animals—raised questions in the minds of some of her readers. Which races of men did Mrs. White have in mind? And did she mean that some races were partially animal? Critics charged her with teaching that Negroes were not members of the human race. But as Uriah Smith pointed out in his defense of *The Visions of Mrs. E. G. White* (1869), a book warmly recommended by James White, such accusations were unfair. The mere possession of some animal blood, Smith argued, did not strip one of humanity, because individuals were human if they had "any of the original Adamic blood in their veins." No one familiar with "the wild Bushmen of Africa, some tribes of the Hottentots, and perhaps the Digger Indians of our own country, &c." could reasonably doubt the validity of Mrs. White's view, wrote Smith. "Moreover, naturalists affirm that the line of demarkation between the human and animal races is lost in confusion. It is impossible, as they affirm, to tell just where the human ends and the animal begins."[21]

Smith also appealed to contemporary science to corroborate Ellen White's statements about the size of humans and animals before the flood. Referring to the recent discovery of what had erroneously been identified as huge human bones, he reported that evidence "is now almost daily coming up fresh from the bosom of the earth—evidence from the discovery of organic remains, sufficient to show beyond a sane doubt, that at some period in the past there existed on this earth a class of gigantic men and animals, in comparison with which the present species are but pigmies."[22] As this quotation illustrates, Adventists did not hesitate to employ the growing authority of science when it served their purposes.

George McCready Price and Flood Geology

Throughout the nineteenth century Seventh-day Adventists could turn to no scientists of their own, except for a few physicians. Thus the appearance of the first Adventist "scientist," George McCready Price, represents a new phase in the history of the denomination's encounter with science. Price was born in eastern Canada in 1870. When his widowed mother joined the Adventist church, he, too, embraced that faith. During the early 1890s Price attended Battle Creek College for two years and subsequently completed a teacher-training course at the provincial normal school in New Brunswick, Canada.[23]

The turn of the century found him serving as principal of a small high school in an isolated part of eastern Canada, where one of his few companions was a Harvard-trained physician. The doctor and the teacher enjoyed discussing scientific matters, and the former almost succeeded in making an evolutionist of his Fundamentalist friend. On at least three occasions Price nearly succumbed to evolution, or at least to what he always considered its basic tenet: the progressive nature of the fossil record. Each time he was saved by sessions of intense prayer—and by reading Mrs. White's book *Patriarchs and Prophets,* which attributed the fossil record to Noah's flood. As a result of this experience, he decided on a scientific career championing what he call the "new catastrophism," in contrast to the old catastrophism of the French naturalist Georges Cuvier.[24]

By 1906 Price was living in southern California and working as a handyman at the Adventists' Loma Linda Sanitarium. That year he published a slim volume entitled *Illogical Geology: The Weakest Point in the Evolution Theory,* in which he confidently offered a $1,000 reward "to any who will, in the face of the facts here presented, show me how to prove that one kind of fossil is older than another." Essentially, he argued that Darwinism rested "logically and historically on the succession of life idea as taught by geology," and that "if this succession of life is not an actual scientific fact, then Darwinism . . . is a most gigantic hoax." Throughout his life Price saved his sharpest barbs for uniformitarian geology, since, in his opinion, "the modern theory of evolution is about 95% due to the geology of [Charles] Lyell and only about 5% to the biology of Darwin."[25]

Readers of Price's book responded in widely divergent, but predictable, ways. The head of a theological seminary in Ohio wrote that he had never read anything clearer and more convincing on the subject, while David Starr Jordan, president of Stanford University and an authority on fossil fishes, told Price frankly that he should not expect "any geologist to take [his work] seriously." Jordan conceded that Price had written "a very clever book" but went on to describe it as "a sort of lawyer's plea, based on scattering mistakes, omissions and exceptions against general truths that anybody familiar with the facts in a general way cannot possibly dispute. It would be just as easy and just as plausible and just as convincing if one should take the facts of European history and attempt to show that all the various events were simultaneous." Jordan's suggestion that Price obtain some "direct contact with problems regarding fossils" penetrated the weakest spot in the creationist's armor: his lack of any formal training or field experience in geology. Price was, however, a voracious reader of geological literature, an armchair scientist who self-consciously minimized the importance of field experience. "It has long been notorious," he once said, "that field naturalists are often mere children when attempting to handle the larger problems of science." Darwin served as an excellent example.[26]

During the next fifteen years Price taught in several Adventist schools and wrote six more books attacking evolution, particularly its geological foundation. Although not unknown in Fundamentalist circles before the early 1920s, he did not begin attracting widespread national attention until then. Shortly after the Fundamentalist controversy entered its anti-evolution phase, Price published his *New Geology,* the most systematic and comprehensive of his two dozen or so books. In it, he restated his "great *law of conformable stratigraphic sequences* . . . by all odds the most important law ever formulated with reference to the order in which the strata occur." According to this law, *"any kind of fossiliferous beds whatever, 'young' or 'old,' may be found occurring conformably on any other fossiliferous beds, 'older' or 'younger.'"* To Price, so-called "deceptive conformatives" (where strata seem to be missing) and "thrust faults" (where the strata are apparently in the wrong order) proved that there was no natural order to the fossil-bearing rocks, all of which he attributed to Noah's flood.[27]

Although *The New Geology* pleased many Fundamentalists, it scarcely improved Price's reputation among practicing scientists. Charles Schuchert of Yale, in reviewing Price's book in the journal *Science,* accused the creationist not only of "harboring a geological nightmare" but of outright dishonesty in appropriating a number of his illustrations from other authors. (In a heated exchange with the editor of *Science* Price protested his innocence and threatened to sue for libel, but the affair was never satisfactorily resolved.) Despite attacks on his views by the scientific establishment, Price's influence among non-Adventist Fundamentalists grew rapidly. By the mid-1920s the editor of *Science* could accurately describe Price as "the principal scientific authority of the Fundamentalists," and Price's byline was appearing with increasing frequency in a broad spectrum of religious periodicals: The *Sunday School Times* and *Moody Monthly* each published about a dozen of his articles, and such diverse journals as *Bibliotheca Sacra, Catholic World, Princeton Theological Review,* and *The Bible Champion* eagerly sought his literary services.[28]

On the eve of the Scopes trial, William Jennings Bryan, the high priest of Fundamentalism, invited Price to assist the prosecution as an expert witness. Although Bryan and Price disagreed over the correct reading of Genesis, Price was a logical choice, being both an acquaintance of Bryan's and the best-known "scientist" in the Fundamentalist camp. Price, however, was teaching at the time in an Adventist college outside London and could not attend the trial. At one point during his notorious examination of Bryan, Clarence Darrow asked if he respected *any* scientist. When Bryan named Price, Darrow scoffed: "You mentioned Price because he is the only human being in the world so far as you know that signs his name as a geologist that believes like you do . . . every scientist in this country knows [that he] is a mountebank and a pretender and not a geologist at all." As we saw in the previous chapter, Price never forgave Bryan for his failure to defend flood geology, a view Bryan never embraced.[29]

While Bryan and Darrow matched wits in Tennessee, Price busily prepared for a showdown in London with Joseph McCabe, a prominent evolutionist and former Jesuit. The debate, held on September 6, 1925, in Queen's Hall, turned into a fiasco. Shortly after the event, Price, a far better writer than a public speaker, complained to a friend in the United States that "during my last 15 minutes I was heckled and interrupted a great deal, and was not permitted to finish as I might have done. At one time, I suppose a thousand people were on their feet at once, yelling and arguing with me or with their next-seat neighbours. It was a lively time." The *New York Times* reported that "interruptions became so frequent that half the audience seemed to be on their feet arguing among themselves. One young woman . . . shouted so determinedly at [Price] that at last he sat down." After such harsh treatment, it is perhaps understandable that Price thought the English were "prejudiced" against him and his views.[30]

His experience at the conservative Victoria Institute in London did little to change his mind. Although the institute awarded him its Langhorne-Orchard Prize in 1925, for an essay entitled "Revelation and Evolution," many members resented the North American's attempt to export the Fundamentalist controversy to England, where science and religion were coexisting in relative harmony. One of the scientists at the institute rebuked Price for attempting "to drive a wedge between Christians and scientists," as had been done in America. The editor of the institute's transactions advised his colleagues that it would be foolish for British Christians to launch a new crusade against evolution simply because Price thought they should. Even a literal reading of the first verses of Genesis, noted the editor, provided ample opportunity for accommodating long-term geological developments.[31]

Late in 1928, as the fires of Fundamentalism burned dim, Price returned to the United States. He continued to preach his "new catastrophism," but came to realize by the late 1930s that he was fighting for a losing cause. Not only was public interest in his crusade fading, but even his own students were beginning to defect. The most traumatic defection was that of Harold W. Clark, who had studied with Price in 1920 and then succeeded him as professor of geology at the Adventists' Pacific Union College in northern California. Later, to Price's annoyance, Clark earned a master's degree in biology from the University of California. In his early scientific writings, such as *Back to Creationism* (1929), Clark followed his mentor closely, but the more he observed, the more he questioned Price's views. Eventually, he broke with Price on three major points: glaciation, stratification, and tectonics.[32]

Beginning with the summer of 1929 Clark devoted his vacations to studying glaciation in the mountains of the West, and the evidence he saw convinced him that, contrary to Price's view, there had indeed been extensive glaciation, that ice had once covered large portions of North America. In the summer of 1938 he visited the oil fields of Oklahoma and northern Texas and received what he later described as a "real shock." For years he had unquestioningly accepted Price's topsy-turvy view of the fossil record, but the order and system he observed in the Southwest convinced him that strata followed a predictable sequence. Other investigators persuaded him that the evidence for overthrusts was "almost incontrovertible."[33]

By 1940 Clark had substituted a non-Adventist text for Price's *New Geology* in his course at Pacific Union College and was describing Price's book as "entirely out of date and inadequate in its handling of its problems." When Price learned of this, he exploded, angrily accusing his former student of suffering from "the modern mental disease of university-itis" and of currying the favor of "tobacco-smoking, Sabbath-breaking, God-defying" evolutionary geologists. Although Clark continued to believe in a literal six-day creation and a universal flood, his acceptance of the geologic column, which depicted the sequence of the fossil-bearing strata and which Price regarded as the primary structure holding up the edifice of evolution, was sufficient evidence for him to conclude that Clark was satanically inspired. Clark repeatedly tried to placate Price, but to no avail. In 1941 Price filed formal heresy charges against Clark with the Adventist governing body on the West Coast. A specially appointed committee of leading Adventists met in San Francisco to investigate Price's charges, but the results proved inconclusive. Nevertheless, Price continued his vitriolic attacks on Clark, which culminated in late 1946 or early 1947 in the publication of a pamphlet entitled *Theories of Satanic Origin*, which unmistakably linked Clark with views that Ellen White had condemned.[34]

Price's conduct in this affair undermined his position as the most respected Adventist spokesman on scientific matters and created sympathy for Clark, who soon offered the denomination a new flood paradigm. To explain paleontological evidence in terms of the biblical record, Clark developed a theory of pre-deluge ecological zones, which held that the various fossil-bearing strata represented the different ecological zones of the antediluvian world. He published this theory in 1946 in a volume called *The New Diluvialism*, the first constructive effort by an Adventist to make sense of the geological record. Until this time Adventist writers, including Price, had devoted most of their energy to poking holes in the prevailing scientific theories.[35]

The 1940s brought modification not only to Price's geological views but to his biological ideas as well, particularly those relating to speciation. In *Genes and Genesis* (1940) Clark took issue with the "extreme creationism" of Price, who insisted on the special creation of all known species (as defined by the Swedish naturalist Carolus Linnaeus). Although Clark believed that no new "kinds" had appeared since the creation, he agreed with Darwin that natural selection had indeed produced many new "species." Four years later another young Adventist biologist, Frank Lewis Marsh, published an even more sophisticated treatment of the species question, in *Evolution, Creation, and Science.* Marsh, a former student of Price's at Emmanuel Missionary College, was the first Adventist to earn a doctoral degree in biology, having received a Ph.D. from the University of Nebraska in 1940. Like Clark, he rejected the Linnaean theory, advocated by Price, that all species had originated by separate creative acts. Zoologists, he pointed out, had identified thousands of species of dry-land animals, yet Adam had been able to name all the newly created animals in one day. Thus it seemed unreasonable to equate the "kinds" of Genesis with the species of the twentieth century. To reduce confusion between the two taxonomic categories, Marsh suggested calling the originally created types *baramins.*[36]

Despite the growing influence of Clark and Marsh, who themselves disagreed on the limits of speciation and the role of amalgamation, Price continued to affect Adventist science well beyond his death in 1963 at the age of ninety-three. During the last decades of his life he worked closely with a small but growing community of Adventists in southern California interested in problems related to creation and evolution. As early as 1936 this group had urged the General Conference of Seventh-day Adventists, the church's international governing body, to sponsor field work in areas such as the Grand Canyon, but the expense of such a program frightened church leaders. Rebuffed, Price and his friends in the Los Angeles area organized the Deluge Geology Society in 1938 to collaborate "in the upbuilding of a positive system of faith-building science." Between 1941 and 1945 they published *The Bulletin of Deluge Geology and Related Sciences,* mailed to over two hundred subscribers. As described by Price, the society consisted of a "a very eminent set of men. . . . In no other part of this round globe could anything like the number of scientifically educated believers in creation and opponents of evolution be assembled, as here in Southern California." Among the active members of the group were a rough-and-ready former evangelist and sometime deputy secretary of state in Arkansas, Benjamin F. Allen, and several physicians, including the socially skilled Molleurus Couperus.[37]

A schism in 1945 between the physicians and Allen resulted in the dissolution of the original group and the creation of the Society for the Study of Natural Science, composed largely of the same membership as the Deluge Geology Society, except for Allen. Until 1948 this organization published *The Forum for the Correlation of Science and the Bible*, edited by Couperus. During its lifetime *The Forum* devoted considerable attention to the age of Earth, with Price (for a time) and Couperus arguing for an Earth "probably older than two billion years" and Clark defending the "ultra-literal view . . . that the matter composing the earth was spoken into existence as the first step in the six-day creation process." Both sides agreed that no life had appeared on Earth before the Edenic creation. By 1947 this organization was dying for lack of interest.[38]

The Geoscience Research Institute

A major breakthrough for Adventist science occurred at a meeting of church leaders in the autumn of 1957. The General Conference Committee, concerned that the church still lacked a single Ph.D. in geology or paleontology, voted "that arrangements be made to send two mature, experienced men of proved loyalty, to take special studies in [geology and paleontology] in qualified institutions for advanced study." The closest the Adventists had to a trained paleontologist was Richard M. Ritland, who had recently received a doctorate from Harvard University in biology and had done considerable work in paleontology and comparative vertebrate anatomy. Within a year the General Conference had selected its candidates for advanced study: the mature and experienced Frank Lewis Marsh and P. Edgar Hare, a young chemistry teacher from Pacific Union College who aspired to become a geochemist. Marsh enrolled at Michigan State University; Hare, at the California Institute of Technology. Together, they represented the Washington-based Research Division of the General Conference Department of Education, which added a third member, Ritland, in 1960.[39]

Before long the Research Division, renamed the Geoscience Research Institute and relocated to Berrien Springs, Michigan, split over methodological issues. Marsh insisted on using the historic Adventist interpretation of the Bible and the writings of Ellen White as the foundation of his scientific investigations. Hare and Ritland, in contrast, expressed a willingness to reinterpret the biblical account of creation and the writings of Mrs. White if the scientific evidence so indicated, an "open-minded" approach their senior colleague regarded as "satanic." Marsh could not understand, for instance, "why [as he thought] both Hare and Ritland and most of the SDA [Seventh-day Adventist] chemistry and physics teachers in our colleges insist on believing in the radioactive timeclocks even after it is known that they place Creation Week hundreds of millions of years ago. The Bible gives us . . . only a few thousand years since Creation Week."[40]

When Hare's research on amino-acid ratios in marine shells yielded a much greater age for the shells than that traditionally accepted by Adventists, he candidly advised the General Conference of the potential problem. He also mentioned the impossibility of working harmoniously with Marsh, who believed "the only value of laboratory research is to corroborate the conclusions one reaches by a study of the Scriptures and Spirit of Prophecy," that is, the writings of Ellen White. The following year President Reuben R. Figuhr notified Hare that he was free to remain with the Carnegie Institution in Washington, where he had been doing postdoctoral research, since the primary purpose of the Berrien Springs institute was to read, write, and study—"looking for inconsistencies in the evolutionary writings that appear"—rather than do original research. Under the circumstances Hare decided to stay at the Carnegie Institution, and his position at the Geoscience Research Institute went to the biologist Harold Coffin, an Ellen White loyalist who shifted the balance of opinion in Marsh's direction.[41]

Through the early 1960s Marsh, the institute's senior member, urged the General Conference to endorse his conservative views. Figuhr, however, felt that the discussion regarding the age of Earth that had gone on during his forty years in the ministry had not really amounted to much, and thus "it wasn't something that [the institute] should put too much time on." In 1964, for both social and intellectual reasons, the General Conference retired Marsh, who attributed his fall to "a no-holds-barred process of indoctrination" carried on by his "open-minded" colleagues. A consolation appointment in the Andrews University department of biology, located adjacent to the institute, seemed to him little better than "banishment into the farthest corner of Siberia."[42]

Ritland, who assumed the directorship of the institute, did indeed prove to be more open-minded than Marsh. Unlike Marsh, who allowed his understanding of the Bible and the writings of Ellen White to determine his science, Ritland believed that God had revealed himself both through nature and through Scripture. Apparent conflicts between the two revelations might just as easily result from misreading the written word as from misinterpreting the book of nature. Using this approach, Ritland prompted many Adventist scientists and not a few administrators to reevaluate their attitudes toward geology and paleontology and to abandon the notion that Noah's flood explained virtually the entire paleontological record. In *A Search for Meaning in Nature* (1970) he emphasized the positive evidence of divine design in the world rather than the negative aspects of modern evolutionary science.[43]

This approach, however, proved too liberal for the administration of Robert H. Pierson, who soon after his election to the presidency of the church in 1966 made his position clear: "In our controversy with proponents of the evolutionary theory," he declared in 1968, "we must keep in clear perspective—the Bible and the Spirit of Prophecy are not on trial." It soon became evident that Ritland had fallen out of favor with the leadership of the church—and that Marsh was now more attuned than he to the pulse of Adventism. In 1971 Ritland, finding it increasingly difficult to function within the constraints imposed by church headquarters, resigned his position as director of the Geoscience Research Institute and joined Marsh in what was becoming an Adventist Siberia, the Andrews University department of biology. The church's brief experiment with open-mindedness thus came to an end.[44]

Under its new director, the conservative physicist Robert H. Brown, the institute quickly swung into line behind the Pierson administration. Adventist scholars who resisted the change soon found themselves without a platform or, worse yet, without a job. A serious problem remained, however. Because Seventh-day Adventists had never adopted a creed, the identification of orthodoxy—and heterodoxy—sometimes proved difficult. To remedy the situation, church leaders, working with the institute's trustees and staff, in 1976 drew up a formal "statement on creation," affirming the denomination's commitment to Ellen White's interpretation of Genesis. The opening paragraph read:

> In harmony with the basic position of the Seventh-day Adventist Church regarding the divine inspiration of the Scriptures, we accept the historical accuracy of the book of Genesis (including chapters 1-11) as providing the only authentic account of the divine creation of this earth and the creation of life upon it in six literal days, of the

fall of man, of the early history of the human race and that of the Noachian Flood of worldwide dimensions.[45]

This distinctive Adventist emphasis on a recent special creation in six literal days and on a fossil-forming flood, long ignored by other Fundamentalists, emerged by the mid-1970s as the intellectual foundation of "scientific creationism" or "creation science." Although proponents usually camouflaged the Adventist origins of their views, the roots of the creation-science movement ran directly back to Price by way of John C. Whitcomb, Jr., and Henry M. Morris, whose book, *The Genesis Flood* (1961), ignited the creationist revival of the late twentieth century. By the last decades of the century Price's flood geology, under the guise of scientific creationism, had become so popular among conservative evangelical Christians that many, ignorant of its history, equated it with traditional biblical creationism.[46]

Adventists, surprisingly, tended to remain on the sidelines while other Christians—from Southern Baptists to Missouri Lutherans—promoted their account of Earth history. Throughout the last quarter of the twentieth century official Adventism, represented most visibly by the General Conference and the Geoscience Research Institute, staunchly defended a hyperliteral reading of the Bible and the writings of Ellen White. However, as Ritland, Hare, and a handful of other liberal Adventists pointed out the scientific futility of trying to squeeze the entire history of life into fewer than 10,000 years, heterodox views became common among Adventist academics. For example, one prominent professor at Loma Linda University, Jack Provonsha, cleverly tried to accommodate fossil evidence of the antiquity of life by attributing many pre-Edenic life forms to Satan's experiments with genetic engineering over "a long period of time" before the appearance of Adam and Eve.[47]

A 1994 survey of the views of Adventist scientists revealed an unprecedented diversity of opinion about creation and the flood. Of the 121 scientists who responded, only 43 percent adhered to the standard belief that "God created live organisms during 6 days less than 10,000 years ago," and only 64.5 percent subscribed to Price's view that "most fossils result[ed] from the world-wide, Bible flood." A sizable 18.2 percent favored theistic evolution, while 3.3 percent took the heretical position that all organisms, including humans, had evolved over immense periods of time "without God's guidance." In response to this troubling news the president of the General Conference, Robert S. Folkenberg, appointed a blue-ribbon panel to study the problem. The church, he insisted, would continue to hold that "God created life on this earth in six literal days, just a few thousand years ago," and that Noah's flood produced most of the fossil record. Clearly, for him, evolutionary biology and geology remained "sciences of satanic origin."[48]

6

CREATION, EVOLUTION, AND HOLY GHOST RELIGION: HOLINESS AND PENTECOSTAL RESPONSES TO DARWINISM

The American Holiness movement arose in the late nineteenth century largely among Methodist followers of John Wesley who sought "entire sanctification" in their lives and a collective return to the primitive ways of the apostolic church. Some of the more radical believers claimed baptism by the Holy Ghost, which allowed them to perform miraculous healings and other wonders. As the movement grew in the 1880s and 1890s—by means of camp meetings, prayer groups, and tracts—Holiness leaders who despaired of reforming Methodism began urging their people to leave their old churches and join independent Holiness groups. By the turn of the century the Church of God (Anderson, Indiana) and the Church of the Nazarene had emerged as two of the most successful of these new denominations.[1]

The symbolic beginning of the Pentecostal movement occurred on the first day of the new century, when Charles Fox Parham, an itinerant Holiness healer in Topeka, Kansas, and a small group of followers first began speaking in tongues (a practice called glossolalia). Inspired by reading about the gifts of the Spirit that had rained on early Christians during the Day of Pentecost, permitting them to talk in unknown languages, Parham had prayed for a "latter rain." Five years later, one of his disciples, William J. Seymour, carried the Pentecostal message to Los Angeles and attracted widespread attention with his Azusa Street revival. The resulting Pentecostal movement came to be most closely identified with the practices of glossolalia and divine healing, from raising the dead to curing the incurable. Promoted in the second half of the twentieth century by such flamboyant evangelists as Oral Roberts and Jimmy Swaggart, Pentecostalism swelled into a mighty international force, converting masses around the globe and promoting so-called charismatic practices among other Christians. By the mid-1990s roughly one fourth of the two billion Christians in the world had embraced the Pentecostal-Charismatic faith.[2]

Shaking the Monkey out of the Cocoanut Tree

In the summer of 1926 the Holiness evangelist Andrew Johnson announced the suspension of his twenty-seven-part attack on "the biological baboon boosters," serialized in the *Pentecostal Herald,* so that he could temporarily return to the camp-meeting circuit. No one in the Holiness community had been agitating more vigorously to "shake the monkey out of the cocoanut tree," as he described his crusade, and he wanted to assure his readers that his "lectures against Darwinism and 'ape to man' evolution" would never eclipse the gospel of salvation. "There is nothing like an old-fashioned, soul-saving revival of Holy Ghost religion," declared the bombastic preacher. "So, let it be distinctly understood that the lectures on Evolution are absolutely secondary to the main line work of intense, soul-saving evangelism to which we have been called and in which we expect to remain."[3]

Johnson's apology epitomizes the ambivalence with which Holiness people greeted the news of their simian ancestry. The notion of kinship with the apes not only offended human pride and seemed to encourage immorality but also contradicted the plain teaching of the Bible and undermined the doctrine of Christian perfection; hence it deserved condemnation. However, unlike their Fundamentalist brothers and sisters from the Calvinist tradition, who stereotypically staked all on the inerrancy of Scripture, conservative Wesleyans, such as those found in the Holiness movement, tended to place experience above exegesis. For them, behavior took precedence over belief. Thus, though they instinctively rejected organic evolution, particularly as it pertained to humans, few of them assigned the issue high priority. During the various outbreaks of antievolution sentiment—following the appearance of the *Origin of Species* in 1859, following the end of World War I, and following the publication in 1961 of *The Genesis Flood* by John C. Whitcomb, Jr., and Henry M. Morris—conservative Wesleyans often proved to be loyal followers, but they generally let others assume the lead.

Because the Wesleyans responded more calmly to Darwinism than did evangelicals of a Reformed bent, historians have tended to overlook their involvement in the evolution controversies. The standard studies of the Holiness movement mention evolution only in passing, if at all, and recent accounts of the Darwinian debates in America, with one notable exception, say little about the Methodists and nothing about the heralds of Holiness.[4] In this chapter I chart the varying, sometimes distinctive, responses to evolution manifest in the Holiness tradition and do so within the context of broader evangelical currents, particularly those flowing from the Reform-minded Fundamentalists and from the tongues-speaking Pentecostals, who carried Holiness teachings to what critics deemed extremes.

Because key elements of my analysis hinge on the changing fortunes of specific interpretations of Genesis 1, it is helpful to review them at the outset. Through the centuries there has been no end to imaginative readings of the first verses of the Bible, but three in particular have competed most successfully for the allegiance of evangelical Christians during the past hundred years. The day-age theory, which allowed the "days" of Moses to span immense geological ages, won the allegiance of such prominent nineteenth-century evangelicals as James Dwight Dana and such leading twentieth-century Fundamentalists as William Jennings Bryan. The ruin-and-restoration, or gap, theory squeezed the geological ages into the period between the original creation "in the beginning" and the subsequent Edenic restoration and circulated in such widely read works as G. H. Pember's *Earth's Earliest Ages* and the *Scofield Reference Bible.* The flood theory, popularized by the Seventh-day Adventist writer George McCready Price during the early twentieth century and by Whitcomb and Morris during the latter part of the century, telescoped earth history into a brief period of about 6,000 years; instead of interpreting the fossils as evidence of past geological ages, it assigned most of them to the one year of Noah's flood. Until the 1960s, the day-age and gap theories dominated evangelical apologetics; after that, the flood theory, packaged as "creation science" or "scientific creationism," attracted so much attention it became virtually synonymous with creationism.[5]

Holiness writers rarely distinguished between organic evolution in general and Darwin's particular theory of natural selection; like most nonscientists, they used the terms evolution and Darwinism interchangeably and primarily to convey a genetic relationship between men and monkeys. "The public, as a rule, has a fairly good understanding of the popular meaning of evolution," explained Johnson. "When the word evolution is pronounced today the public can almost hear the bark falling from the cocoanut trees and catch the faint echoes of the chattering of the chimpanzee. The people know that the so-called scientific doctrine of evolution somehow or other conveys the idea that man came from the monkey."[6]

In his presidential address to the American Society of Naturalists in 1891 the Methodist geologist William North Rice reflected on "the revolution of opinion upon the subject of Evolution" during the past quarter-century. In 1867, the year in which as a twenty-one-year-old he had received the first doctorate in geology awarded by Yale University, few American scientists, including Rice himself, had yet embraced Charles Darwin's recently published theory. But just twenty-five years later the "fingers of one hand" sufficed "to count all anti-evolutionists who are competent to have an opinion on the subject." Indeed, only one prominent scientist in all of North America, the geologist John William Dawson of Montreal, remained in the antievolutionist camp. Rice, like most of his scientific colleagues, remained a theist, and taught for years at the Methodists' Wesleyan University. Yet he denied that Genesis taught history or that it could be harmonized with the findings of modern geology.[7]

While Holiness revivalists were preaching the gospel of entire sanctification, their more liberal brethren, especially in the North, were coming to terms with the theological implications of Darwinism and higher criticism of the Bible, which treated Scripture like any other ancient text. We still have no comprehensive study of Wesleyan responses to evolution, but most Victorian Methodists seem not to have shared Rice's enthusiasm for the teachings of Darwin. As E. Brooks Holifield has shown for the English church, they especially "feared that Darwin had defamed the moral character of man." Decades ago the historian Windsor Hall Roberts argued that American Methodists neither "took a leading part in the fight against evolution" nor played "a great part in the adjustment of Christianity to evolution." He attributed this lack of involvement to their preoccupation with church growth and to the fact that the Methodists "drew the bulk of their people from the less intellectual elements of the population who, although opposed to evolution, were more or less inarticulate." Even Roberts, however, acknowledged that, during the last decades of the nineteenth century, many of the scientific and theological leaders of Methodism accepted some version of organic evolution.[8]

No one contributed more to the scientific education of nineteenth-century Methodists than the devout but controversial geologist Alexander

Winchell, who, in the words of David N. Livingstone, served as a "scientific go-between" for his church. Although Vanderbilt University, a Methodist school, dismissed him for teaching that human history antedated Adam and Eve, the progressive editor of the *Methodist Quarterly Review,* Daniel D. Whedon, provided him with a platform from which to educate fellow believers regarding the merits of what had come to be known colloquially as Darwinism. By the early 1890s evolution had won such widespread support among the clergy that, according to Rice, no self-respecting seminary could be regarded as complete "without a course of lectures on the consistency of Evolution with theistic philosophy." Now and then, he conceded, "some theological Rip van Winkle attempts the old Sinaitic thunders in denunciation of the essential atheism of evolution; but his utterances are regarded by his brethren in the church, not with sympathy, but with amusement or mortification."[9]

The most vocal of the Methodist Rip van Winkles was Luther T. Townsend, sometime professor at Boston Theological Seminary, who wrote such works as *Evolution or Creation, Adam and Eve,* and the mass-circulated pamphlet that informed Americans about the alleged demise of Darwinism in Europe, *Collapse of Evolution.* Although Townsend defended the first chapters of Genesis as "a simple, straight-forward narrative of the facts as they actually occurred," he readily accommodated the findings of geology by resorting to the gap theory. This allowed him to interpose "vast geological epochs" between the original creation of heaven and Earth and the Edenic restoration, at which time "in six literal days, and in the order given in the Bible, the Creator brought the world out of the chaos of the glacial wreck, made it habitable, created modern flora and fauna and gave them life and power to propagate themselves until the end of time." Townsend's effectiveness as a critic of evolution led to his being considered as an appropriate candidate to write on the subject for *The Fundamentals* (1910–1915), the series of booklets that launched the Fundamentalist movement. But the editors eventually selected the more scientifically qualified George Frederick Wright, a Congregational cleric-geologist.[10]

During the formative years of the Holiness movement, as it split off from Methodism and metamorphosed into separate institutional churches, its leading ministers paid scant attention to scientific matters, choosing rather to preach "straight Holy Ghost religion." They believed that the Bible taught spiritual, not scientific, truths; yet they rejoiced whenever the findings of modern science seemed to corroborate scriptural events. By and large, they had little trouble accepting the development of the solar system from a nebula, the progression of geological ages, or even a geographically limited flood. However, when it came to human evolution, they drew the line.[11]

Typical in this regard was the evangelist W. B. Godbey, a prolific writer fascinated by scientific themes. In such works as *Man the Climax of Creation, Regenerated Earth,* and *Bible Astronomy,* he displayed no antipathy toward either the nebular hypothesis or the notion of geological epochs, and he went out of his way to disabuse naive Christians of "the popular idea" that Earth was young and that the "days" of Genesis were literal. "Our people much need light, to simplify and elucidate these controverted problems," he wrote, "because infidels take advantage of them and sidetrack our people into clouds of darkness." To obtain the maximum possible time, he conflated the gap and day-age interpretations of Genesis 1, arguing that God, in preparing the earth for humans, had taken "millions of ages"—all prior to the six days of the Edenic creation, which were themselves "demiurgic, creative periods," not solar days. He showed no toleration, however, for human evolution: "The Darwinian infidelity teaches . . . that we were evolved out of apes, gorillas, orang-outangs. We know this is false simply because it flatly contradicts the Bible. . . . This settles the matter forever."[12]

The idea of human evolution seemed to strip the doctrine of original sin of conventional meaning by reducing it to animal inheritance—a point readily conceded by some theistic evolutionists. At a national congress of the Methodist Episcopal Church, held in 1898, the Methodist-reared biologist E. G. Conklin scandalized Holiness delegates by declaring that "evolution explains the moral unrest of human kind as due to the conflict between the animal and the spiritual, the beastly and the heavenly; it explains original sin as brute inheritance and the fall as a conscious yielding of the higher to the lower nature." According to a writer for the *Christian Witness and Advocate of Bible Holiness,* the professor had committed a "monumental blunder. . . . This extreme view of evolution we must reject because it collides with and is diametrically opposed to the biblical account of man's creation, his fall and utter helplessness to recover himself without divine aid."[13]

Because of the bearing of evolution on human history, the Free Methodists of Illinois passed a resolution in 1903 forbidding their preachers to teach "theistic evolution." One Free Methodist minister, Alexander Beers, president of Seattle Pacific College, offered an incendiary solution to the problem: "Take all the books and make a pyramid as high as the snowy crest of Mt. Rainier, take all of J. D. R[ockefeller]'s oil and pour it over them and set fire to it. Take all the promises of the blessed old Book and read them during the conflagration, then join in singing, 'Praise God from whom all blessings flow.' "[14] Such statements, though few in number, suggest that Darwinism made little headway among turn-of-the-century Holiness people. Evolution and Holy Ghost religion simply did not mix.

The Crusade against Evolution

In the wake of World War I, concern about the moral consequences of Darwinism, especially its impact on the youth of America, led to an unprecedented crusade that culminated in the passage of several statewide laws banning the teaching of evolution. As the evolution question polarized American society and modernist biblical scholarship seeped into the mainline churches, Holiness people found it increasingly difficult to stand on the sidelines. Although they continued to stress their distinctive brand of experiential evangelicalism, many of them contributed positively to what Paul Merritt Bassett has called the "fundamentalist leavening" of the conservative Wesleyan tradition.[15]

During the heyday of the antievolution movement in the 1920s many Holiness bodies joined the Fundamentalists in censuring Darwinism. In 1923 delegates to the General Holiness Convention in Indianapolis expressed sympathy for the Fundamentalists in the battle against modernism, while noting that soundness of doctrine was insufficient for salvation. On the question of evolution, they affirmed that "the Holiness Movement *uniformly* believes that man is the product of God's immediate creation." Because Darwinism seemed to be "Satan's greatest and most subtle form of attack upon the faith of the world in the fact of the supernatural, the deity of Christ, the inspiration of the Scriptures and the instantaneity of salvation," they earnestly protested the teaching of "this unproven hypothesis" in both public and religious schools. Nevertheless, they cautioned against investing too much time and energy in the antievolution crusade:

> We certainly do not mean that our brethren are to forsake the preaching of salvation from sin as has been our custom, but that there shall be warning given, and information and light thrown upon the subject by those who are capable of doing so, so that the people shall not

> go unwarned. But for the rank and file of the movement—those whose position and scholarship do not warrant an effective discussion of the subject in public, we advise that they shall press the battle for immediate salvation, and through prayer and testimonies and their ministry both private and public seek to precipitate upon this land and every where a genuine revival of old time Holy Ghost scriptural holiness.[16]

Warnings against the evils of evolution appeared in publications put out by groups ranging from the relatively prosperous Nazarenes to the socially marginal Pentecostals, whose claims to speaking in tongues reminded one Holiness writer of the equally unsubstantiated claims of the evolutionists.[17] In the mid-1920s the Church of the Nazarene, the largest of the Holiness denominations, with over 63,000 members in 1926, launched its offensive against Darwinism with such works as E. P. Ellyson's *Is Man an Animal?* and Basil W. Miller and U. E. Harding's "*Cunningly Devised Fables,*" which described evolution as "the sputum of hell" and "the van guard of the war on God's revealed Word." Written almost entirely by Miller (Harding's name was added for window dressing because he was an evangelist), the latter book minced no words in denouncing the "double-tongued, false-hearted, scabby-souled . . . imps of damnation" who destroyed faith in the biblical doctrine of creation by teaching that humans developed from "the slimy ooze."[18]

The Free Methodist Church, with some 36,000 members, continued to follow in the footsteps of Alexander Beers and B. T. Roberts, its founder, in opposing evolution. During the 1920s the *Free Methodist* carried a steady stream of antievolution articles, including one by George McCready Price, the father of flood geology.[19] The *Wesleyan Methodist,* official organ of the Wesleyan Methodist Church, with just under 22,000 members, endorsed the ban against evolution in public schools and hawked books by William Jennings Bryan. A popular book by the Wesleyan Methodist evangelist George D. Watson, *God's First Words,* concluded that "the whole teaching of evolution is absolutely atheistic and a denial of God and of creation and leads to every form of falsehood in doctrine, in life, in morals and everything else."[20]

The Pilgrim Holiness Church, with approximately 15,000 members, ran occasional pieces against evolution in the *Pilgrim Holiness Advocate,* and one of its ministers, B. H. Shadduck, produced some of the most tasteless tracts of the time. In a series of crudely illustrated booklets with such titles as *Jocko-Homo: The Heaven-Bound King of the Zoo, Puddle to Paradise,* and *Rastus Augustus Explains Evolution,* a story in black dialect about "a pompous old colored man" who picks up the rudiments of evolution while working as a college janitor, Shadduck skewered evolutionists with sarcasm and humor. He justified his heavy-handed approach on the grounds that most other antievolution books were "too lengthy for tired working people and busy students" and were too scholarly to be "readily grasped by common folk." "I regret," he wrote disingenuously, "that I know of no way to heave a brick at the old squaw [evolution] and seem perfectly polite to the enamored bystanders." Frustrated by the "pussy-footing" of the typical evolution fighter, he decided the time had come to "SHELL THE WOODS."[21]

The various Pentecostal churches, though instinctively opposed to evolution, devoted noticeably less attention to the question than Holiness folk and far less attention than Fundamentalists. As Vinson Synan has observed, the most outspoken defamers of Darwin during the Scopes trial "were not the 'Holy Rollers on Shinbone Ridge,' as H. L. Mencken implied, but the Presbyterians, Baptists and Methodists in the Courthouse." The Pentecostals' low level of participation in the antievolution crusade may have reflected their general indifference to formal education and, hence, to what American youth were learning in school, or it may have resulted from their intense focus on receiving the gifts of the Spirit rather than decoding the Word. The Assemblies of God, the largest of the Pentecostal denominations, said relatively little about Darwinism in the *Pentecostal Evangel,* and the few antievolution pieces that did appear simply followed the Fundamentalist line.[22] The Pentecostal Holiness Church virtually ignored the Darwinian debate, though in the early 1930s George Floyd Taylor, a onetime general superintendent and founding editor of the *Pentecostal Holiness Advocate,* published a twenty-eight-part "Exposition on Genesis 1-3." In this novel work, he interpreted the first day of creation as a twenty-four-hour period but speculated that "long periods of time" separated the subsequent days, which "may have themselves been long durations of time."[23] Aimee Semple McPherson, the flamboyant founder of the International Church of the Foursquare Gospel, said more about evolution than most Pentecostal preachers, probably because as a high school student her own faith had been temporarily shaken by the theory. In the early 1930s, she publicly debated the question with the atheist Charles Lee Smith.[24]

The low importance many Pentecostals attached to the creation-evolution controversies is epitomized by an anonymously written article that appeared in the *Church of God Evangel,* a magazine published by the

Church of God in Cleveland, Tennessee. Because of the atypical subject matter it contained, the editor added an introductory explanation:

> Some of our good Pentecostal folks are compelled for business reasons to live where they are constantly coming in contact with higher criticism. Here are the observations of a Church of God Brother who has come in direct contact with all these things, and yet stands true. Do you think you could stand true if you didn't get to be with many people who believed in the whole Bible, the Church of God, and the other things so dear to you? This brother did.

The author himself granted that scientists possessed a logical advantage over "us Pentecostal people" in dealing with such topics as evolution and that science was "a fine thing" for agriculture and industry. Yet science could not prove that "we don't speak in tongues," and it had not yet demonstrated the evolutionary origins of humans. If it ever did, the writer professed to find no grounds for concern:

> The way I look at it now I don't think it matters much whether we were a part of a wet clay bank and we were piled up a pile of mud, and made living men with the faculties of a human being or whether we were anthropoid apes hanging by our tails in some primeval jungle. . . . To have had either ancestry would be found to demonstrate the greatness of God in bringing out of it a man like you.[25]

It is difficult to imagine such sentiments appearing in a Fundamentalist tract of the times.

The epicenter of antievolution activity in the Holiness world was located not at the headquarters of the newly formed churches but in Wilmore, Kentucky, home of the interdenominational Asbury College and Theological Seminary and the influential *Pentecostal Herald.* At the head of all three of these institutions stood the imposing figure of Henry Clay Morrison, the acknowledged leader of the Holiness movement within the Methodist Episcopal Church, South, and an orator of Bryanesque proportions. In a front-page story on October 19, 1921, Morrison alerted readers of the *Herald* to "the coming conflict between the conservatives and the liberals with reference to the inspiration and trustworthiness of the Bible":

> the first shots have been fired, the hosts are gathering and the battle is on. We welcome the conflict most heartily. We shall watch its progress with intense interest and THE HERALD will, without hesitation, train its guns on those men and teachings who, if permitted to go unrebuked, will destroy the faith of the people in the inspiration of the Bible. . . . Let every faithful soldier of the cross draw his sword and hasten to the firing line in the coming conflict between saving faith and destructive unbelief.

His call did not go unheard nor his promise unkept. For the next seven years or so, he and a squad of sharpshooters from Wilmore fired round after round at evolution and related heresies. Besides Morrison, the Wilmore creationists included John Paul, the vice president of Asbury College; George W. Ridout, its professor of apologetics; L. L. Pickett, an associate of Morrison's in the Methodist Episcopal Church, South; and Andrew Johnson, president of the Wilmore-based Fundamentalist Association and one of the most dogged anti-Darwinists in Southern Methodism.[26]

Morrison also launched numerous salvos into the evolutionist camp from the Louisville-based Pentecostal Publishing Company. This press, which he controlled, not only issued a stream of antievolution booklets by such Holiness worthies as Godbey, Shadduck, Pickett, and Ridout but also reprinted Townsend's creationist classic, *Collapse of Evolution,* and at least three of William Jennings Bryan's works. Neither the Presbyterian Bryan nor the Methodist Townsend joined the Holiness cause, but both were widely praised and published in Holiness circles, and by the time of Bryan's death in 1925, shortly after the Scopes trial, Holiness people had adopted him as one of their own heroes.[27]

The Holiness creationists justified their sometimes harsh attacks on evolution in terms of the damage it seemed to be doing to faith and morals. "Nothing has done so much to destroy evangelical saving faith among the people as the evolutionary hypothesis," declared Morrison. "It makes animals out of human beings; then why not seduce them, destroy them with drink, deceive, cheat and wrong them in any and every way." Morrison's logic was graphically applied in a popular story for young people, *The Not-Ashamed Club,* written by an Assemblies of God author. Two young men, raised as Christians, had lost their faith, taken to heroin, and committed murder. At their trial one of them traced the course of their downfall:

> at school we were taught and made to actually believe that men were not created by God as the Bible says, but that by what they call evolution men and animals developed out of the same original life cell. . . . This made it plain to us that whatever good times we were to ever have must be had in this life. It also took from us the feeling in which we had both been brought up that there is something sacred about human life, different from the life of a beast. . . . The sacredness of human life was taken from us as a necessary logical result of the statements of evolution.

The youths felt no remorse at the time of the killing, but they knew others would disapprove of their act because, as one of them testified, "people who are uneducated and do not understand the origin of species, feel that it is a bad thing indeed to kill anyone." Under the circumstances, the jury could not bring itself to convict the youths. As the foreman explained, "we old heads that allow evolution to be taught in our schools, and allow heroin to be shaken under the noses of our boys, are the ones to blame."[28]

Indeed, young people seemed particularly vulnerable to the wiles of evolution, and Holiness literature abounded with sad tales of students led astray by godless professors—often at Methodist schools. "The great Methodist church has not an outstanding university or college true to the fundamentals," claimed the Nazarenes Miller and Harding.

> Run the list, Vanderbilt, Northwestern, Syracuse, University of Southern California, Ohio Wesleyan, Boston University, and it is found that liberalism in its rankest form is often taught . . . only a few of the weaker colleges of the North, some of the smaller ones of the South, and those institutions fostered by the Church of the Nazarene, the Free Methodist Church, the Wesleyan Methodist Church and other holiness groups and the numerous Bible schools are true to the fundamentals.

A survey of selected midwestern colleges, conducted in 1919, offers corroboration for this generalization. In responding to the question "What is the teaching of your scientific and sociological chairs respecting the doctrine of evolution?" the half dozen or so Methodist presidents in the poll tended to endorse theistic evolution. In contrast, the heads of the two Holiness schools who filled out the questionnaire expressed a preference for special creation. "They are opposed to it," the president of Greenville College, a Free Methodist school in Illinois, replied tersely. "We limit the question simply to the development of individual and specific forms giving more emphasis to creation and epochs than ordinary evolutionists admit," answered the president of Central Holiness University (later John Fletcher College) in Iowa.[29]

Although Morrison estimated that "nine-tenths of the preachers and people of Southern Methodism" remained true to the Word of God, even Southern Methodists were inclined to watch the antievolution contest rather than participate in it. "Orthodoxy of opinion is secondary to the orthodoxy of life," explained a Southern Methodist writer in the *North Carolina Christian Advocate.* "Not what a man thinks about religion but a personal religious experience is primary." When the Southern Methodist Educational Association went so far as to deplore "all legislation that would interfere with the proper teaching of scientific subjects in American schools and colleges," Morrison immediately protested to the bishop.[30]

It is easy to detect the not-so-hidden agenda behind the Holiness criticism of Methodist schools. Conservative Holiness leaders wanted not only to drive evolutionists from Methodist colleges but also to build up their own institutional empires, and they eagerly seized evolution as a handy club with which to attack the churches and schools of more liberal Methodists. As Paul Merritt Bassett has observed, the antievolution crusade of the 1920s occurred at the very time that Holiness churches were beginning to establish their own liberal arts colleges. In order to win needed financial and moral backing, "the Movement followed a characteristic strategy: whenever broad support is needed, yield as much as necessary to even the most conservative folks, for the more liberal people will still go along. . . . This strategy guaranteed that the Movement would be kept astir about evolutionary theory for the agents for the colleges went about assuring everyone that their schools would be up-to-date as to the issues." Even Bible schools used creation as bait to recruit students. For example, the Carolina Bible Training school in Highfalls, North Carolina, billed itself as "a safe place to study," where "evolution with all of its relatives is branded as offsprings of the devil." The back cover of the school's magazine fittingly carried an advertisement for Shadduck's *Jocko-Homo: The Heaven-Bound King of the Zoo.*[31]

To stop the advance of evolution, Morrison and his associates recommended two courses of action. First, they called for the firing of all teachers and preachers in religious institutions who advocated evolution. "Heresy," thundered Johnson, "must be made odious and heretics must be punished—not by death or imprisonment, but by the official condemnation of their theories and ex-communication in case they refuse to recant." If evolutionists desired to teach their insidious theory, Morrison thought it only fair that they "build their own schools, endow their own colleges, establish their own theological seminaries and furnish the plaster [of] Paris at their own cost to manufacture monkeys and apes from various and sundry bones they may happen to pick up, to prove that man's ancient ancestors were mere animals." Second, the Wilmore creationists supported legislation that would outlaw the teaching of evolution in public schools. If that initiative failed, as they feared it would, then they favored a tax strike. "We can easily conceive of two or three hundred thousand Kentuckians refusing to pay taxes to furnish salaries to conceited professors who think it quite smart to ridicule Revelation and teach their pupils that their ancestors were apes," wrote Morrison.[32]

Because of their opposition to evolution, Holiness creationists ran the risk of appearing to be antiscience. In listing their priorities, they always ranked science below salvation. Holiness people, advised the Nazarene theologian Ellyson, should not spend so much time in the laboratories of the natural sciences that they neglected "the laboratory of Pneumatology [i.e., the study of the Spirit] and the development of the higher spiritual and immortal powers." As Asbury's Professor Ridout pointed out, religion was acquired not "in the laboratory, but at the mourner's bench." To dispel the impression of anti-intellectualism, Holiness apologists repeatedly professed their love of "all true scientific discovery." In a series entitled "Methodism and Modern Thought" written for the *Pentecostal Herald,* Andrew Johnson began defensively by asserting that "Methodism is not opposed to all modern things. It is not opposed to modern civilization, to modern inventions, to modern improvements, to modern scientific discoveries, to modern conveniences, to modern methods." One observer, noting recent developments in communication and transportation, concluded that "constructive science is being rapidly perfected with the sole object of revealing the glory of God, and a more rapid proclamation of the Gospel of Christ to a lost world."[33]

Morrison especially feared that thoughtless disparagements of science might discredit the Holiness cause. In "An Open Letter to a Young Preacher," Morrison warned not to let

> the modernists or skeptics of any brand provoke you to any sort of unwise or sweeping statement against any branch of science or scien-

> tists in general. I have heard some very loud and boisterous declarations made by indignant ministers with much pointing of the fist that would raise questions in the minds of the thoughtful. . . . All intelligent ministers of the Gospel and devout Christians are ready to receive gladly any scientific fact that has been proven; in other words, *truth established.* . . . It is your privilege and duty to discern between truth established and the mere theories of men.

Such populist constructions of science stemmed less from philosophical predilections than from a commonsense reading of ready reference works. "But what is science?" asked Pickett rhetorically in *God or the Guessers.* "To the dictionaries!" he answered. There one learned that science was "certified and classified knowledge." Because evolution remained an uncertified *theory,* it fell outside the domain of true science. By defining evolution as nonscience, Holiness critics, like their Fundamentalist friends, no doubt reduced the dissonance created by their rejection of widely held expert opinion.[34]

To qualify as science in the minds of these creationists, evolution required empirical demonstration. Johnson, who prided himself on his knowledge of biology, established two tests for securing scientific status. First, the evolutionist "must either manufacture life in the chemical retort or laboratory, or he must furnish a verified instance of spontaneous generation in the laboratory of nature." Second, he must give at least a "single instance where one clearly-defined species has changed into another species." Then, and only then, could he claim evolution as a scientific fact. In the meantime, Johnson would go along with his sometime collaborator Pickett in labeling it " 'Guessology,' for it is the unscientific science of poor guessing."[35]

In dismissing biological evolution as nonscience, the creationists often went out of their way to defend the scientific legitimacy of geological development. "There is no divergence, discord, gap or gulf between Genesis and Geology," insisted Johnson. "The great break is between Geology and Evolution." Morrison concurred, carefully distinguishing between the acceptable theory of geological ages and the unacceptable theory of Darwinian evolution. Most Holiness writers easily accommodated the findings of geology by postulating a gap between the original creation of Earth, described in Genesis 1:1, and the much more recent Edenic creation, described in the rest of the chapter. "This earth may have been inhabited by man and beast in preadamic days and then by some process unknown to us, it may have been made void and useless like a worn out garment," speculated one commentator. "This could be true and yet not one word of the Mosaic account of the creation would be discredited, because we know that the six days' work of creation had nothing to do with the creation of the earth and its being made void, but only with the reshaping and the placing on the same both plant and animal life." I have discovered no Holiness writer from this period insisting on a young Earth or invoking Noah's flood to account for the fossil record, as did the Seventh-day Adventists, and at least Johnson entertained the possibility of nonsolar "days."[36]

Among the numerous Holiness commentators on evolution, no one displayed greater sophistication than Samuel A. Steel, a University of Virginia–trained Southern Methodist pastor and Chautauqua lecturer whose byline occasionally appeared in the *Pentecostal Herald.* Although he sometimes sounded like a Fundamentalist—"if we accept the theory of evolution, we may as well send our Bibles to the scrap heap"—he ventured considerably further than most of his brethren in embracing the findings of modern science and biblical scholarship. In *The Modern Theory of the Bible,* ostensibly a reply to the higher critics, not only did he describe the first chapter of Genesis as "history and science poetically expressed," but he also emphasized the correspondence of the Genesis narrative with modern science when the "days" of the former were taken to represent indefinite periods of time. Both science and revelation, he argued, taught that "the world was made by a gradual process." In a statement widely quoted by antievolutionists, the naturalist Joseph LeConte had defined evolution as "a continuous, progressive change from the lower to the higher forms of life according to certain laws and by means of resident forces." Instead of ritualistically denouncing this notion as unbiblical, Steel argued that "Moses only reveals the origin of these 'resident forces' in the will of God." This was unmistakably the rhetoric of theistic evolution, as was his comment that "evolution, properly understood, does not separate nature from God, but brings God into nature, so that what we call natural law is but the mode of the divine activity." As a creationist, Steel insisted on only "three great outstanding 'breaks' in the scientific continuity of nature: the origin of matter; the origin of life; and the origin of mind."[37]

At least some Holiness writers blamed the creation-evolution controversy on the failure of scientists to honor the traditional boundary separating science from religion. One contributor to the *Pentecostal Herald* accused scientists of provoking the current conflict by imperialistically invading the territory of the theologians:

> It has not been long since theology was a department for the theologian, with his Bible, and science was a department for the scientist with his test tubes and microscope; but rapidly we are seeing the way change. The scientist has demanded the right to invade the field of the theologian, and is now telling him the ways of God through test tubes and microscopes. We are perfectly willing for the scientist to go on with his legitimate work. He is needed, his work is important, the world eagerly waits for his many possible discoveries and inventions in the way of making human life longer and easier; but let the scientist keep his hands off of our theology and our Bible, as he demands that the theologian shall keep his hands off of his science with its laboratory. The two departments of science and theology are separate and distinct, and it is up to us as Christian men and women to see to it that they remain so.

The message was clear: Without well-patrolled borders, there could be no true peace.[38]

In support of their claim that evolution was unscientific, creationists compiled extensive rosters of scientists who allegedly shared their point of view. Shortly after the turn of the century, Luther T. Townsend assembled one of the earliest—and most frequently cribbed—lists in order to prove that "the most thorough scholars, the world's ablest philosophers and scientists, with few exceptions, are not supporters, but assailants of evolution." In addition to such distinguished but deceased scientists as Louis Agassiz, Arnold Guyot, and John William Dawson, he cited the English physiologist Lionel S. Beale and the German pathologist Rudolf Virchow, both of whom had expressed doubts about the evidence for human evolution, as well as a dozen or so lesser lights, including a "Dr. Etheridge, of the British Museum, one of England's most famous experts in fossilology," and "Professor Fleischmann, of Erlangen, one of the several recent converts to anti-Darwinism." Robert Etheridge was a little-known assistant keeper of geology at the British Museum, who spent his last years in obscurity in Australia. During the early decades of the twentieth century, the German zoologist Albert Fleischmann was the only living biologist of repute to contribute to the creationist critique of evolution.[39]

Throughout the antievolution crusade of the 1920s, Holiness writers who wished to discredit Darwinism scientifically continued to appeal to Townsend's authorities—and occasionally added a new scientific expert to the list. In *God or the Guessers,* Pickett identified Townsend himself, a biblical scholar, as "a real scientist." And in his series "Evolution Outlawed by Science," Johnson promoted two scientific amateurs to world-class standing. He described the Catholic priest George Barry O'Toole, author of *The Case against Evolution,* as "the great orthodox modern biologist," and he elevated the Seventh-day Adventist schoolteacher George McCready Price to the ranks of the "greatest living geologists."[40]

In spite of the much-heralded collapse of Darwinism, the antievolution movement in America failed to recruit a single prominent scientist. One of the few creationists with any graduate training in biology was the Holiness professor S. James Bole, who taught successively at Missouri Wesleyan College, the Fundamentalist Wheaton College in Illinois, the Holiness John Fletcher College in Iowa, and Taylor University, a nondenominational Holiness school in Indiana. A graduate of the University of Michigan, Bole went on to earn a master's degree in education from the University of Illinois and spent six additional years in Champaign-Urbana pursuing a Ph.D. in pomology (fruit culture) in the school of agriculture. As a college student, Bole had accepted biological evolution; but during his years at Illinois, he experienced a "Damascus Road" conversion at a Paul Rader evangelistic service and consequently rejected evolution and higher criticism in favor of the Mosaic story of creation and biblical inerrancy. Shortly after he moved to Wheaton, an acquaintance at the nearby Moody Bible Institute in Chicago urged him to use his experience and expertise to write a book against evolution. The resulting work, which advocated the ruin-and-restoration interpretation of Genesis 1 and a nonuniversal deluge, appeared in book form in 1926 under the title *The Modern Triangle.* The Pentecostal Publishing Company distributed it, and Holiness leaders such as Ridout warmly recommended it. In 1932 Wheaton College dismissed Bole for alleged social improprieties, and thereafter he remained on the sidelines while other creationists, such as the Adventist Price, began building an institutional infrastructure in the 1930s.[41]

The Creationist Revival

For years Price had dreamed of having his own creationist journal and society, but he refused to associate himself with anything that smacked of Shadduck's "'Jocko-Homo,' 'Puddle-to-Paradise' style of argument." Largely to support a dignified publication, he joined with Dudley Joseph Whitney, a onetime disciple of the Pentecostal faith healer Alexander Dowie, in forming a society to sponsor a magazine soberly called *The Creationist.* Thus was born the Religion and Science Association, apparently the first antievolution organization in America aimed at resolving scientific and hermeneutical problems rather than restricting the teaching of evolution. Price and Whitney's first task was to recruit a prestigious panel of officers. Another ex-Dowieite, L. Allen Higley, who held a Ph.D. in chemistry from the University of Chicago and taught science at Wheaton College, came on board as president. He in turn suggested inviting Jay Benton Kenyon, of Asbury College, to serve as a member of the five-man board of directors, but the idea generated little enthusiasm. Whitney considered asking Shadduck but worried about his tactless style: "He has a good head, though I did not like his rough stuff." Eventually the group settled on a couple of Lutherans and another Seventh-day Adventist.[42]

Despite all that Whitney and Higley shared in common, the two former Pentecostals soon found themselves in bitter disagreement on the subject of Genesis 1. Whitney, following Price, insisted on a recent creation and a geologically significant flood, while Higley, an unbending gap theorist, assigned most geological history to a pre-Edenic period. Their controversy quickly engulfed the entire association and helped to bring about its rapid demise. In reaction to this unpleasant turn of events, Price and some Seventh-day Adventist friends in the Los Angeles area organized the exclusive Deluge Geology Society in 1938. For nearly a decade, this organization provided an intellectual home for like-minded creationists around the country. Among the early recruits was William J. Tinkle, a Ph.D.-credentialed biologist at La Verne College, just east of Los Angeles, who believed in geological catastrophism but resisted attributing all the fossil-bearing strata to a single flood. Nevertheless, he attended society meetings when possible and occasionally published in the *Bulletin of Deluge Geology.* Though a member of the Church of the Brethren, Tinkle moved in Holiness circles and later served on the faculties of both Taylor University and Anderson College in Indiana.[43]

In 1941 five evangelical scientists concerned about the quality of the Christian witness on science and religion organized the American Scientific Affiliation. Within a decade, membership had grown to 220, over half of whom represented Baptist, Presbyterian, and Mennonite congregations. Only nine members identified themselves as Methodists, three as Free Methodists, two as Nazarenes, and two as belonging to the Christian and Missionary Alliance. Despite its tiny Wesleyan contingent, the association held its second annual convention, in 1947, on the campus of Taylor University, where Tinkle was teaching. Cecil B. Hamann, a biologist from Asbury, presented a paper, and Paul E. Parker, from the Wesleyan Methodist Marion College, chaired a session. During the early years of the ASA, no one urged a harder line against evolution than Tinkle, who co-wrote the chapter "Biology and Creation" in the association's handbook, *Modern Science and Christian Faith,* published in 1948. But by the early 1960s, a coalition of theistic evolutionists and progressive creationists had captured the association.[44]

In 1961, Whitcomb and Morris published their landmark book, *The Genesis Flood,* which triggered the creationist revival of the late twentieth century and more than any other work brought Price's flood geology to the attention of American evangelicals. Two years later, at a joint meeting of the ASA and the Evangelical Theological Society at Asbury College, dissident creationists allied with Whitcomb and Morris withdrew to Midland, Michigan, to form their own Creation Research Society. Of the eighteen original members of "the inner-core steering committee," only Tinkle, who had now moved on to Anderson College, operated by the Church of God (Anderson), claimed any connection with the Wesleyan-Holiness tradition.[45]

Because of their unprecedented success in pushing evangelicals to choose between young- and old-Earth histories, after the early 1960s the flood geologists increasingly dictated the terms of the debate over origins. But although one survey in the early 1960s showed the Nazarenes (80 percent) and Assemblies of God (91 percent) trailing only Seventh-day Adventists (94 percent) in the percentage of members opposed to human evolution, flood geology made its greatest gains within the doctrinally based Fundamentalist churches, not in the experientially oriented Holiness and Pentecostal communions. Holiness magazines paid relatively little attention to *The Genesis Flood,* and when they did notice it, they did so without enthusiasm. James F. Gregory, editor of the *Free Methodist,* predicted that the book would "become an authority for those who believe that creation was a unique act of God, and that the flood of Noah's time was universal." S. Hugh Paine, a physics professor at the Wesleyan Methodist Houghton College, praised Whitcomb and Morris for their "objective and scholarly manner" and particularly for their arguments against "the uniformitarian hypothesis—an unprovable thesis which by its very nature forbids any literal acceptance of the Genesis record." He hoped that their critique of uniformitarianism would prevail "in the ranks of Evangelical scholarship," but he feared that "deep-seated prejudices" would prevent most readers from accepting its radical thesis. Like his old Wheaton professor L. Allen Higley, Paine himself favored what he called the "Gap-Flood" model of earth history, which assigned most of the fossil-bearing rocks to a pre-Edenic flood rather than to Noah's deluge. This scheme possessed the advantage of allowing creationists to accept the evidence for "the apparent great age of the earth, the apparent tenure of ancient populations of pre-Adamic animals, and the apparent pre-Adamic hominid paleontology."[46]

By this time virtually all of the major Holiness denominations had set aside their early reservations about higher education and had invested heavily in liberal arts colleges. In part to provide adequate training for premedical students, these schools had gradually developed programs in the biological sciences. As early as the 1920s, for example, Asbury College designated David William Nankivel as "professor of biology." For years he offered a course called Problems in Evolution, in which he (perhaps unthinkingly) promised to refute both "facts and theories." In the mid-1930s the college asked Jay Benton Kenyon, a theologically trained "professor of chemistry," to assume Nankivel's title. It was not until after World War II, in 1946, that Asbury hired its first adequately trained biologist, Cecil B. Hamann, an alumnus of Taylor who had gone on to obtain a Ph.D. in biology from Purdue University. A self-styled progressive creationist, he allowed the period of creation to extend over thousands, even millions, of years, occasionally punctuated by special creative acts. Until his retirement in the 1980s he effectively defended this middle-of-the-road position against theistic evolutionists on the one flank and flood geologists on the other. At the end of the decade, Asbury College still had no theistic evolutionists on its science faculty, but only one professor, a computer specialist, showed any sympathy for flood geology.[47]

Other Holiness colleges sometimes hired theistic evolutionists to teach biology, but these professors generally maintained a low profile to avoid becoming the targets of a more conservative constituency. By and large, Holiness scientists shied away from doctrinaire flood geology, as did their associates at Pentecostal schools. In 1972, the president of the Assemblies of God Evangel College invited the Morris-led Institute for Creation Research to use his campus in Springfield, Missouri, for the first Summer

Institute on Scientific Creationism. Upon arriving, the flood geologists discovered no supporters among the science faculty. Later, when Turner Collins, a plant pathologist, co-offered a seminar on science and religion, he used texts and guests that were anathema to flood geologists. About the same time, L. Duane Thurman, a botanist from Berkeley on the faculty of Oral Roberts University, wrote *How to Think about Evolution*, in which he irenically evaluated a number of acceptable interpretations of Genesis. Myrtle M. Fleming, who earned a Ph.D. in biology from the University of Georgia and who built up the science programs at both Lee College, operated by the Church of God (Cleveland), and Emmanuel College, run by the Pentecostal Holiness Church, warned Pentecostal theologians not to be taken in by claims for a young Earth. In writing and teaching about origins, she said, "great care should be taken to distinguish between facts and theory, original works and philosophers' thinking."[48]

Theologians in the Wesleyan-Holiness tradition, like their scientific colleagues, tended to stick to the familiar gap and day-age readings of Genesis 1 rather than switch to the novel flood theory. *The Wesleyan Bible Commentary*, the first such work produced by evangelical Wesleyans after the beginning of the creationist revival, ignored flood geology altogether. The Wesleyan Methodist author of the section on Genesis, Lee Haines, read the first chapters of the Bible through scientifically tinted lenses, keeping a constant eye on what science "required." Because both the gap and day-age schemes made "room for the geologic ages and for the fossils of extinct, prehistoric beasts," he judged both acceptable, though the idea of a ruin and restoration suffered from being "almost entirely based on supposition." Haines no longer regarded evolution as the "bugaboo" of biblical scholarship, but he thought the Bible restricted biological variation to the Genesis "kinds."[49]

Two years later the Nazarenes brought out the *Beacon Bible Commentary*, which began with an explication of Genesis by George Herbert Livingston, a Free Methodist. Livingston, too, believed that the phrase "after its kind" ruled out macroevolution, but beyond endorsing the defense of indefinite "days" by the Nazarene theologian H. Orton Wiley, he did not indicate how he thought the creation story should be read. He showed no awareness at all of the current debate over earth history. He twice cited *The Genesis Flood* as a factual authority but never mentioned its hermeneutic significance. In one jarring passage on Noah's flood, he artlessly juxtaposed references to the Whitcomb-Morris book and Bernard Ramm's *Christian View of Science and Scripture* without even hinting that the two works represented opposite schools of thought on the topic.[50]

Eugene F. Carpenter, in the article "Cosmology" for Charles W. Carter's *Contemporary Wesleyan Theology* (1983), faulted Wesleyan scholars for paying inadequate attention to the topic. He attributed this oversight to two factors: "First, Wesley's theology is centered around practical pastoral issues and, second, Wesleyan theology has been intensely concerned with man; it is, in a good sense, man-centered." The few existing treatments of the topic left Carpenter less than satisfied. In his opinion, Wiley had spent "too much time relating science and Christianity, even drawing in the outdated gap theory of Genesis 2 [*sic*] in order to harmonize natural science and Scripture." Haines had been too enamored of modern science. Livingston had insufficiently emphasized "the moral and religious nature of the cosmos." Such interpretations contrasted with the example of John Wesley and Adam Clarke, the most revered biblical scholars in the Methodist tradition, who had stressed "the moral and religious dimensions of creation."[51]

Except for the theologians, few Holiness writers turned to the exegetical writings of the eighteenth-century Wesley and the early nineteenth-century Clarke for enlightenment in deciphering the meaning of Genesis. Although Wesley had been inclined to take the biblical cosmogony literally and dismissed the developmental speculations of the French naturalist Buffon as "utterly inconsistent both with reason and Scripture," he warned against reading too much science into the creation story. In commenting on the first chapter of Genesis, he urged readers to remember that the Scriptures "were written not to gratify our curiosity, but to lead us to God." Because Wesley accepted the so-called chain of being, which organized all of creation into a graded series, "from dead earth, through fossils, vegetables, animals, to man," some early-twentieth-century admirers tried to claim him as a proto-evolutionist. This provoked a spirited defense of Wesley's orthodoxy. "These wicked and insane evolutionists must stop their base slandering of John Wesley," demanded one outraged Methodist cleric. The Holiness community seems to have paid little attention to this tempest in a teapot, although in 1925 the *Pentecostal Herald* did reprint one defense of Wesley as a creationist.[52]

It is hazardous to offer any generalizations about recent Wesleyan Holiness attitudes toward origins, except to observe that flood geology made little headway among scientists and theologians in the community and that it created less of a stir in Wesleyan than in Reformed circles. The literature of the past few decades contains articles both for and against flood geology, sometimes running side by side. For example, in a 1964 issue of *Light and Life*, published by the Free Methodists, a retired pastor approvingly quoted Henry M. Morris in support of a universal Noachian flood. A few pages later, a scientifically trained layman censured the unnamed "founder of the Creation Research Society" for his "devious" arguments and suggested that those who sought "scientific evidence to prove the Genesis account of Creation" lacked faith in divine revelation. As for himself, he professed not to be troubled by the question of evolution. Just how many fellow believers shared his inclination to hold the issue in abeyance is impossible to determine.[53]

On the surface Pentecostals seemed equally undisturbed by the rising tide of flood geology. Many of the most vocal and visible preachers saw no reason to abandon the ruin-and-restoration theory found in the trusted *Scofield Reference Bible* for the "creation science" of Fundamentalist flood geologists, who tended to dismiss speaking in tongues and healings as satanic delusions. The endorsement of the gap theory in *Dake's Annotated Reference Bible,* an immensely influential work that began displacing *Scofield* among conservative Pentecostals in the early 1960s, no doubt contributed significantly to its continuing popularity. Prepared by Finis Jennings Dake, a Pentecostal evangelist with a checkered legal and denominational history, *Dake's,* as it was commonly called, gave eighteen "proofs" of the existence of a pre-Adamic world destroyed by a pre-Edenic catastrophe he called Lucifer's flood. Such Pentecostal stalwarts as Jimmy Swaggart, Lester Sumrall, Kenneth E. Hagin, and Gordon Lindsay also remained loyal to a pre-Edenic gap. Some commentators, such as Dake, restricted the days of creation to twenty-four hours each, but a few, including Lindsay, followed the Pentecostal pioneer Charles F. Parham in expanding the "re-creative days" to six periods of 1,000 years.[54]

Below the surface, however, "creation science" seemed to be winning considerable support among rank-and-file Pentecostals. The Assemblies of God, though taking no official position on the issue, provided John C. Whitcomb, Jr., with an early forum in the *Pentecostal Evangel* and continued into the 1970s and 1980s to print positive notices about the Creation Research Society and the Institute for Creation Research. Flood geology also appeared in works by influential Church of God (Cleveland) and Church of God of Prophecy writers, as well as by various independent charismatic groups. Winkie Pratney of Last Days Ministries, who "gave up a promising career in research chemistry" to become an evangelist, increasingly promoted flood geology in the successive editions of his slick, mass-circulated tract *Creation or Evolution?* By the 1980s, Gordon Lindsay's son, Dennis, was teaching a course on Scientific Creationism at his late father's Christ for the Nations Institute in Dallas.[55]

What, then, can be said about Wesleyan-Holiness responses to creation and evolution? During the century spanning the period from the Holiness revival of the 1890s to the late 1990s, most advocates of Holy Ghost religion who expressed themselves on the subject have stood solidly in the creationist camp, resolutely refusing to become what one Holiness wit called "Evolutes." Some believers, especially in recent decades, have accepted evolution, but the virtual silence of these theistic evolutionists testifies loudly to the prevailing creationist ethos. Though mostly staunch creationists, Holiness people have remained reluctant antievolutionists. Moreover, instead of fretting like some Fundamentalists over the lack of consensus on the correct reading of Genesis, they have rarely worried about the diversity of creationist views in their midst. Their relative equanimity reflects the Wesleyan emphasis on the heart over the mind, on behavior over belief. Even the militant evolution fighters of the 1920s, who matched their Fundamentalist brothers and sisters in vociferous denunciations of Darwinism, felt the need to justify their involvement in such distracting activities. "Twenty-five years ago," observed H. C. Morrison in the early 1920s, "it seemed to be the special work of the holiness movement to proclaim full salvation to believers, but now their sphere of labor has become greatly enlarged, and they must contend for all the fundamental doctrines of the Bible, yea, for the inspiration of the Bible itself." Even so, he and his cohorts never forgot that defending creationism remained "absolutely secondary" to what Andrew Johnson called "the main line work of intense, soul-saving evangelism."[56]

NATURALISTS IN THE NATIONAL ACADEMY OF SCIENCES, 1863–1900

The biographical notes below provide personal data for each of the eighty naturalists and identify their most explicit statements on evolution and the most helpful biographical accounts that I found. Additional biographical information for all but James Hall, J. D. Whitney, and Horatio C Wood can be found in the *Biographical Memoirs* of the National Academy of Sciences. Names in brackets identify people who kindly assisted me in my quest for pertinent facts. Disciplines in brackets following the NAS "committee" to which members were elected, if known, indicate the field of greatest activity.

Agassiz, Alexander (1835–1910) *Family:* Eldest child and only son in a family of three children; father, a professor. *Education:* A.B., Harvard, 1855; B.S., Lawrence Scientific School, Harvard, 1857, 1862. *Employment:* Harvard Museum of Comparative Zoology, 1860–1910. *NAS:* Elected 1866, zoology. *Religion:* Agnostic. *Evolution:* Leaned toward evolution but remained skeptical of grand claims and Darwin's emphasis on natural selection. *Publication:* "Paleontological and Embryological Development," *American Journal of Science* 120 (1880): 294–302, 375–389. *Biographies:* George R. Agassiz, ed., *Letters and Recollections of Alexander Agassiz with a Sketch of His Life and Work* (Boston: Houghton Mifflin, 1913); Mary P. Winsor, *Reading the Shape of Nature: Comparative Zoology at the Agassiz Museum* (Chicago: University of Chicago Press, 1991), pp. 148–155. [Edward Lurie]

Agassiz, Jean Louis Randolphe (1807–1873) *Family:* Eldest surviving child; father, a Protestant minister. *Education:* Ph.D., Erlangen, 1829; M.D., Munich, 1830. *Employment:* Harvard, 1847–1873. *NAS:* Elected 1863, zoology. *Religion:* Unitarian. *Evolution:* Antievolutionist. *Publications:* "Prof. Agassiz on the Origin of Species," *American Journal of Science* 80 (1860): 142–154; "Evolution and Permanence of Type," *Atlantic Monthly* 33 (1874): 92–101; *Essay on Classification*, ed. Edward Lurie (Cambridge: Harvard University Press, 1962). *Biographies:*

Jules Marcou, ed., *Life, Letters, and Work of Louis Agassiz,* 2 vols. (New York: Macmillan, 1896); Edward Lurie, *Louis Agassiz: A Life in Science* (Chicago: University of Chicago Press, 1960); Mary Pickard Winsor, "Louis Agassiz and the Species Question," *Studies in History of Biology* 3 (1979): 89–117; Paul J. Morris, "Louis Agassiz's Arguments against Darwinism in His Additions to the French Translation of the *Essay on Classification,*" *Journal of the History of Biology* 30 (1997): 121–134.

Allen, Joel Asaph (1838–1921) *Family:* Eldest of three sons; father, a carpenter and farmer. *Education:* Attended Lawrence Scientific School, Harvard, 1862–1865. *Employment:* Harvard Museum of Comparative Zoology, 1867–1885; American Museum of Natural History, 1885–1921. *NAS:* Elected 1876, biology [ornithology]. *Religion:* Unknown, though he was raised a strict Congregationalist. *Evolution:* Neo-Lamarckian who tended to emphasize the "direct modifying influence of environment." *Publication: Selected Works of Joel Asaph Allen,* with an introduction by Keir B. Sterling (New York: Arno Press, 1974), which includes Allen's "Autobiographical Notes." [Keir B. Sterling]

Baird, Spencer Fullerton (1823–1887) *Family:* The third of seven children; father, a lawyer. *Education:* B.A., Dickinson, 1840; M.A., Dickinson, 1843. *Employment:* Smithsonian Institution, 1850–1887. *NAS:* Elected 1864, zoology [ornithology]. *Religion:* Attended church but was privately an agnostic. *Evolution:* Probably accepted evolution but seems never to have expressed his views publicly. *Biography:* E. F. Rivinus and E. M. Youssef, *Spencer Baird of the Smithsonian* (Washington: Smithsonian Institution Press, 1992).

Beecher, Charles Emerson (1856–1904) *Family:* First son in a family of five children; father, a banker. *Education:* B.S., Michigan, 1878; Ph.D., Yale, 1889. *Employment:* New York State Museum, 1878–1888; Peabody Museum, Yale, 1888–1904. *NAS:* Elected 1899, geology and paleontology. *Religion:* Unknown. *Evolution:* Neo-Lamarckian described as a "leader of the Hyatt School." *Publication:* "The Origin and Significance of Spines: A Study in Evolution," *American Journal of Science,* 156 (1898): 1–20. *Biography:* Robert T. Jackson, "Charles Emerson Beecher," *American Naturalist* 38 (1904): 407–426.

Billings, John Shaw (1838–1913) *Family:* First surviving child and only son in a family of two children; father, an unsuccessful farmer and shopkeeper. *Education:* A.B., Miami, 1857; A.M., Miami, 1860; M.D., Medical College of Ohio, 1860. *Employment:* U.S. Surgeon General's Office, 1862–1895; New York Public Library, 1896–1913. *NAS:* Elected 1883, anthropology [bacteriology and hygiene]. *Religion:* Rejected Presbyterianism for non-Christian theism, perhaps even agnosticism. *Evolution:* Probable evolutionist who personally liked Darwin and Huxley, but his precise views remain unknown. *Biographies:* Fielding H. Garrison, *John Shaw Billings: A Memoir* (New York: G. P. Putnam's Sons, 1915); Carleton B. Chapman, *Order Out of Chaos: John Shaw Billings and America's Coming of Age* (Boston: Boston Medical Library, 1994). [James H. Cassedy]

Boas, Franz (1858–1942) *Family:* A laterborn child in a family of six children and the only son to survive childhood; father, a moderately successful merchant. *Education:* Ph.D., Kiel, 1881. *Employment:* Clark, 1889–1892; Columbian Exposition, 1892–1894; American Museum of Natural History, 1895–1905; Columbia, 1896–1937. *NAS:* Elected 1900, anthropology. *Religion:* Secular Jew, self-described materialist. *Evolution:* Sympathetic to the inheritance of acquired characteristics and to the role of "inherent forces . . . that control the development of the animate world," but mentioned biological evolution only guardedly in his writings and refrained from discussing it in his seminars. *Writings:* "The Relation of Darwin to Anthropology," undated ms., Boas Papers, Library of the American Philosophical Society, Philadelphia; "Franz Boas," in *I Believe,* ed. Clifton Fadiman (New York: Simon and Schuster, 1939), pp. 19–29; George W. Stocking, Jr., *The Shaping of American Anthropology, 1883–1911: A Franz Boas Reader* (New York: Basic Books, 1974). *Biography:* Marshall Hyatt, *Franz Boas, Social Activist: The Dynamics of Ethnicity* (Westport, CT: Greenwood Press, 1990). [Herbert S. Lewis, George W. Stocking, Jr.]

Bowditch, Henry Pickering (1840–1911) *Family:* Son of a well-to-do merchant; birth order unknown. *Education:* A.B., Harvard, 1861; A.M., Harvard, 1866; M.D., Harvard Medical School, 1868; additional study in Europe. *Employment:* Harvard Medical School, 1871–1906. *NAS:* Elected 1887, zoology [physiology]. *Religion:* Unknown. *Evolution:* Unknown. *Biography:* W. Bruce Fye, *The Development of American Physiology: Scientific Medicine in the Nineteenth Century* (Baltimore: Johns Hopkins University Press, 1987), pp. 92–128. [W. Bruce Fye]

Brewer, William Henry (1828–1910) *Family:* Son of a farmer; birth order unknown. *Education:* Ph.B., Sheffield Scientific School, Yale, 1852; A.M., Yale, 1859; additional study in Germany. *Employment:* California State Geological Survey, 1860–1864; Sheffield Scientific School, Yale, 1864–1903. *NAS:* Elected 1880, biology. *Religion:* Unknown, though probably Protestant. *Evolution:* Neo-Lamarckian who defended the inheritance of acquired characters in the face of Weismann's criticism. *Publication:* "On the Hereditary Transmission of Acquired Characters," *Agricultural Science* 6 (1892): 103–107, 153–156, 249–254, 345–348. [Nancy G. Slack]

Brooks, William Keith (1848–1908) *Family:* The second of four sons; father, a merchant. *Education:* B.A., Williams, 1870; Ph.D., Harvard, 1875. *Employment:*

Johns Hopkins, 1876–1908. *NAS:* Elected 1884, biology [zoology]. *Religion:* Agnostic who retained a belief in the possibility of immortality. *Evolution:* A self-described "ardent disciple of Darwin," whose position on variation and inheritance lay "midway between Darwin's view of the origin of variation and the Lamarckian view." *Publications: The Law of Heredity: A Study of the Cause of Variation, and the Origin of Living Organisms* (Baltimore: J. Murphy, 1883); "Lamarck and Lyell: A Short Way with Lamarckians," *Natural Science* 8 (1896): 89–93; "Lyell and Lamarckism: A Rejoinder," ibid. 9 (1896): 115–119. *Biography:* Keith R. Benson, "William Keith Brooks (1848–1908): A Case Study in Morphology and the Development of American Biology," Ph.D. diss., Oregon State University, 1979. [Keith R. Benson]

Brown-Séquard, Charles-Edouard (1817–1894) *Family:* Only child of an American naval officer, who died before his son's birth, and a French woman from Mauritius. *Education:* M.D., Paris, 1846. *Employment:* Medical College of Virginia, 1854–1855; Harvard Medical School, 1864–1867; Collège de France, 1878–1894; medical practice in Philadelphia, New York, Boston, London, and Paris. *NAS:* Elected 1868 [physiology]. *Religion:* Definitely a Protestant, probably an Anglican. *Evolution:* A probable evolutionist, whose research provided the strongest evidence for the inheritance of acquired characters. *Biographies:* J. M. D. Olmsted, *Charles-Edouard Brown-Séquard: A Nineteenth Century Neurologist and Endocrinologist* (Baltimore: Johns Hopkins University Press, 1946); Michael J. Aminoff, *Brown-Séquard: A Visionary of Science* (New York: Raven Press, 1993).

Clark, Henry James (1826–1873) *Family:* Son of a Swedenborgian minister; birth order unknown. *Education:* A.B., University of the City of New York, 1848; B.S., Lawrence Scientific School, Harvard, 1854. *Employment:* Lawrence Scientific School, Harvard, 1860–1865; Pennsylvania State College, 1866–1869; Kentucky, 1869–1872; Massachusetts Agricultural College, 1872–1873. *NAS:* Elected 1872 [histology and microscopy]. *Religion:* Unknown. *Evolution:* Subscribed to the evolutionary views of Richard Owen. [Clark A. Elliott]

Cook, George Hammell (1818–1889) *Family:* Third son in a family of seven children; father, a farmer of moderate means. *Education:* C.E., Rensselaer, 1839; B.N.S., Rensselaer, 1840. *Employment:* Rutgers, 1853–1889. *NAS:* Elected 1887 [geology]. *Religion:* Active Presbyterian. *Evolution:* No known views on the subject. *Biography:* Jean Wilson Sidar, *George Hammell Cook: A Life in Agriculture and Geology* (New Brunswick, NJ: Rutgers University Press, 1976).

Cope, Edward Drinker (1840–1897) *Family:* Eldest of four children, including a step-brother; father, a wealthy gentleman farmer and philanthropist. *Educa-*

tion: Studied at Pennsylvania and for several months in Europe. *Employment:* Haverford, 1864–1867; Pennsylvania, 1886–1897; served on various western surveys. *NAS:* Elected 1872, biology, geology, and geography. *Religion:* Resigned from the Society of Friends in 1878 but remained a theist, a self-described Unitarian. *Evolution:* A leader of the neo-Lamarckian school who accepted evolution in the 1860s. *Publications:* "On the Origin of Genera," *Proceedings of the Academy of Natural Sciences of Philadelphia* 20 (1868): 242–300; *On the Hypothesis of Evolution: Physical and Metaphysical* (New Haven: Charles C. Chatfield, 1871); "A Review of the Modern Doctrine of Evolution," *American Naturalist* 14 (1880): 166–179, 260–271; *The Origin of the Fittest: Essays on Evolution* (New York: Macmillan, 1887); *The Primary Factors of Organic Evolution* (Chicago: Open Court, 1896). *Biographies:* Henry Fairfield Osborn, *Cope, Master Naturalist: The Life and Letters of Edward Drinker Cope* (Princeton: Princeton University Press, 1931); Peter J. Bowler, "Edward Drinker Cope and the Changing Structure of Evolutionary Theory," *Isis* 68 (1977): 249–265.

Coues, Elliott (1842–1899) *Family:* Second son of a second marriage; father, merchant turned U.S. patent officer. *Education:* A.B., Columbian, 1861; M.D., Columbian, 1863. *Employment:* U.S. Army, 1862–1881; Columbian, 1877–1886. *NAS:* Elected 1877, biology [ornithology]. *Religion:* Embraced Theosophy in 1884 and subsequently founded the Gnostic Theosophical Society of Washington, from which he was expelled in 1889. *Evolution:* Accepted "descent with modification," which he attributed to changing environmental conditions. *Publication: Field Ornithology* (Salem, MA: Naturalists' Agency, 1874). *Biography:* Paul Russell Cutright and Michael J. Broadhead, *Elliott Coues: Naturalist and Frontier Historian* (Urbana: University of Illinois Press, 1981).

Dall, William Healey (1845–1927) *Family:* First child and only son in a family with apparently only two children; father, largely absent missionary to India. *Education:* No formal collegiate training. *Employment:* U.S. Coast and Geodetic Survey, 1871–1884; U.S. Geological Survey, 1884–1925. *NAS:* Elected 1897, geology and paleontology. *Religion:* Unknown, though his father was a Unitarian minister. *Evolution:* Advocated saltatory evolution. *Publication:* "On a Provisional Hypothesis of Saltatory Evolution," *American Naturalist* 11 (1877): 135–137. [Phillip Drennon Thomas]

Dalton, John Call, Jr. (1825–1889) *Family:* Eldest of four sons; father, a prominent physician. *Education:* A.B., Harvard, 1844; M.D., Harvard Medical School, 1847; additional study in Paris. *Employment:* College of Physicians and Surgeons, New York City, 1855–1889. *NAS:* Elected 1864, anatomy and physiology. *Religion:* Unknown, though his funeral was held in an Episcopal church. *Evolution:* Unknown. *Biography:* W. Bruce Fye, *The Development of American*

Physiology: Scientific Medicine in the Nineteenth Century (Baltimore: Johns Hopkins University Press, 1987), pp. 15–53. [Thomas Gariepy]

Dana, James Dwight (1813–1895) *Family:* Eldest of twelve children; father, a successful merchant. *Education:* A.B., Yale, 1833. *Employment:* Yale, 1856–1890. *NAS:* Elected 1863, zoology [geology]. *Religion:* Congregationalist. *Evolution:* Came to accept theistic evolution in the 1870s but continued to insist that "a creative act" was necessary for the origin of humans; leaned more toward neo-Lamarckian than Darwinian mechanisms. *Publications: Manual of Geology* (Philadelphia: Theodore Bliss, 1863) and subsequent editions; "Views of Evolution," *Independent* 32 (January 8, 1880): 2–3. *Biographies:* Daniel C. Gilman, *The Life of James Dwight Dana* (New York: Harper and Brothers, 1899); William F. Sanford, Jr., "Dana and Darwinism," *Journal of the History of Ideas* 26 (1965): 531–546; Michael L. Prendergast, "James Dwight Dana: The Life and Thought of an American Scientist," Ph.D. diss., University of California, Los Angeles, 1978.

Dutton, Clarence Edward (1841–1921) *Family:* Son of a shoemaker; birth order unknown. *Education:* B.A., Yale, 1860; briefly attended Yale Theological Seminary. *Employment:* U.S. Army, 1862–1901; U.S. Geological Survey, 1879–1890. *NAS:* Elected 1884, geology and paleontology. *Religion:* Despite his flirtation with the ministry, he became an agnostic. *Evolution:* Expressed support for Cope's views. *Biographies:* Wallace E. Stegner, "Clarence Edward Dutton, Geologist and Man of Letters," Ph.D. diss., University of Iowa, 1935; Stephen J. Pyne, *Dutton's Point: An Intellectual History of the Grand Canyon* (Grand Canyon Natural History Association, 1982). [Stephen J. Pyne]

Emmons, Samuel Franklin (1841–1911) *Family:* Third son and fifth child; father, a successful merchant. *Education:* A.B., Harvard, 1861; A.M., Harvard, 1866; additional study in France and Germany. *Employment:* Geological Exploration of the Fortieth Parallel, 1867–1877; U.S. Geological Survey, 1879–1911. *NAS:* Elected 1892, geology and paleontology. *Religion:* Unknown. *Evolution:* Unknown. [Hatten S. Yoder, Jr.]

Engelmann, George (1809–1884) *Family:* Eldest of thirteen children; father, a university-trained schoolmaster. *Education:* M.D., Wurzburg, 1831. *Employment:* Medical practice in St. Louis, Missouri. *NAS:* Elected 1863, botany. *Religion:* From a family of Reformed ministers; attended university with the assistance of a Reformed congregation. *Evolution:* Despite some claims to the contrary, there is no evidence that he ever accepted evolution; in fact, he generally opposed speculations and theories. *Biographies:* Walter B. Hendrickson, "An Illinois Scientist Defends Darwinism: A Case Study in the Diffusion of Scientific Theory," *Transactions of the Illinois State Academy of Science* 65 (1972): 25–29; Michael Long, "George Engelmann and the Lure of Frontier Science," *Missouri Historical Review* 89 (1994–95): 251–268. [Phillip Drennon Thomas, Patricia Timberlake]

Farlow, William Gilson (1844–1919) *Family:* Son of a successful businessman and state legislator; birth order unknown. *Education:* A.B., Harvard, 1866; M.D., Harvard Medical School, 1870; additional study in Europe. *Employment:* Harvard, 1874–1896. *NAS:* Elected 1879, biology [botany]. *Religion:* Unknown, though probably Protestant. *Evolution:* Accepted evolution in the 1860s but left little record of his views, except for his rejection of Asa Gray's version of theistic evolution. *Publications:* "The Conception of Species as Affected by Recent Investigations on Fungi," *American Naturalist* 32 (1898): 675–696; "The Change from the Old to the New Botany in the United States," *Science*, n.s., 37 (1913): 79–86. [Paul D. Peterson]

Gabb, William More (1839–1879) *Family:* Son of a salesman; birth order unknown. *Education:* B.A., Central High School of Philadelphia, 1857. *Employment:* California State Geological Survey, 1862–1868; geological surveys in Santo Domingo and Costa Rica, 1869–1876. *NAS:* Elected 1876 [geology and paleontology]. *Religion:* Unknown. *Evolution:* Unknown. [Elizabeth N. Shor]

Gilbert, Grove Karl (1843–1918) *Family:* Third child and second son; father, a successful portrait painter. *Education:* A.B., Rochester, 1862. *Employment:* Ward Natural Science Establishment, 1863–1868; Geological Survey of Ohio, 1869–1870; Wheeler's Geographical Survey West of the 100th Meridian, 1871–1874; Powell's Geographical and Topographical Survey of the Colorado River of the West, 1874–1879; U.S. Geological Survey, 1879–1918. *NAS:* Elected 1883, geology and paleontology. *Religion:* Unknown. *Evolution:* Accepted the evolution of species but left little record of his views. *Publication:* "The Work of the International Congress of Geologists," *American Journal of Science* 134 (1887): 430–451. *Biography:* Stephen J. Pyne, *Grove Karl Gilbert: Great Engine of Research* (Austin: University of Texas Press, 1980).

Gill, Theodore Nicholas (1837–1914) *Family:* Birth order and father's occupation unknown. *Education:* Attended Wagner Free Institute of Science, Philadelphia. *Employment:* Columbian, 1860–1910; Smithsonian Institution, 1861–1866; Library of Congress, 1866–1874. *NAS:* Elected 1873, biology [zoology]. *Religion:* Unknown; rejected his father's entreaties to become a minister. *Evolution:* Enthusiastic evolutionist inclined to accept neo-Lamarckism but skeptical about the inheritance of acquired characters. *Publications:* "On the Relations of the Orders of Mammals," *Proceedings of the American Association for the Advancement* *of Science* 19 (1870): 267–270; "The Doctrine of Darwin," *Proceedings of the Biological Society of Washington* 1 (1880–1882): 47–55; "Edward Drinker Cope, Naturalist—A Chapter in the History of Science," *Science,* n.s., 6 (1897): 225–243; "Systematic Zoology: Its Progress and Purpose," ibid. 26 (1907): 489–505. [Victor G. Springer]

Goodale, George Lincoln (1839–1923) *Family:* Son of a pharmacist and state official; birth order unknown. *Education:* A.B., Amherst, 1860; M.D., Harvard Medical School, 1863. *Employment:* Bowdoin, 1869–1872; Harvard, 1872–1909. *NAS:* Elected 1890, biology [botany]. *Religion:* Unknown. *Evolution:* Believed "that plants can be profoundly modified by different external conditions, and that these modifications tend to persist." *Publication:* Review of *The Survival of the Unlike,* by L. H. Bailey, *American Journal of Science,* 153 (1897): 77.

Goode, George Brown (1851–1896) *Family:* Only child; father, a merchant. *Education:* A.B., Wesleyan, 1870; additional study at Harvard. *Employment:* Wesleyan, 1871–1877; U.S. Fish Commission and U.S. National Museum, 1873–1877; Smithsonian Institution, 1877–1896. *NAS:* Elected 1888, biology [ichthyology]. *Religion:* Unknown. *Evolution:* Praised the *Origin of Species* and incorporated evolution into the exhibits at the Smithsonian Institution but left little written record about his views. *Publication: The Origins of Natural Science in America: The Essays of George Brown Goode,* ed. Sally Gregory Kohlstedt (Washington, D.C.: Smithsonian Institution Press, 1991). [Sally Gregory Kohlstedt, Michael J. Lacey]

Gould, Augustus Addison (1805–1866) *Family:* Eldest of five surviving children; father, a musician. *Education:* A.B., Harvard, 1825; M.D., Harvard Medical School, 1830. *Employment:* medical practice in Boston. *NAS:* Elected 1863, zoology. *Religion:* Devout Baptist. *Evolution:* Rejected evolution in 1860. *Publication:* Editor's introduction to *The Naturalist's Library* (Boston: Crosby, Nichols, Lee, 1860). [Clark A. Elliott]

Gray, Asa (1810–1888) *Family:* Eldest of eight children; father, a farmer and tanner. *Education:* M.D., Fairfield College of Physicians and Surgeons, 1831. *Employment:* Harvard, 1842–1888. *NAS:* Elected 1863, botany. *Religion:* Active Congregationalist. *Evolution:* Described himself as "one who is scientifically, and in his own fashion, a Darwinian"; believed that God guided the course of evolution. *Publications: Darwiniana: Essays and Reviews Pertaining to Darwinism* (New York: D. Appleton, 1876); *Natural Science and Religion* (New York: Charles Scribner's Sons, 1880). *Biography:* A. Hunter Dupree, *Asa Gray, 1810–1888* (Cambridge: Harvard University Press, 1959).

Guyot, Arnold Henri (1807–1884) *Family:* One of twelve children, born eleven years after his parents' marriage; father's occupation unknown. *Education:* Ph.D., Berlin, 1835. *Employment:* Princeton, 1854–1884. *NAS:* Elected 1863, astronomy, geography, and geodesy. *Religion:* A devout Presbyterian who as a youth studied for the ministry. *Evolution:* Although widely regarded as an antievolutionist, in his later years, according to his friend James Dwight Dana, "he was led to accept, though with some reservation, the doctrine of evolution through natural causes." *Publications:* "Cosmogony and the Bible; or, The Biblical Account of Creation in the Light of Modern Science," in *History, Essays, Orations, and Other Documents of the Sixth General Conference of the Evangelical Alliance, Held in New York, October 2–12, 1873,* ed. Philip Schaff and S. Irenaeus Prime (New York: Harper & Brothers, 1874), pp. 276–287, 319; *Creation; or, The Biblical Cosmogony in the Light of Modern Science* (New York: Charles Scribner's Sons, 1884). *Biography:* Ronald L. Numbers, *Creation by Natural Law: Laplace's Nebular Hypothesis in American Thought* (Seattle: University of Washington Press, 1977), pp. 91–100.

Hague, Arnold (1840–1917) *Family:* Laterborn son of a Baptist minister. *Education:* Ph.B., Sheffield Scientific School, Yale, 1863; additional study in Germany. *Employment:* Geological Exploration of the Fortieth Parallel, 1867–1877; government geologist in Guatemala, 1877–1878; government geologist in China, 1878–1880; U.S. Geological Survey, 1880–1917? *NAS:* Elected 1885, geology and paleontology. *Religion:* Unknown. *Evolution:* Unknown, though he was generally indifferent to theories. [Clifford M. Nelson]

Haldeman, Samuel Steman (1812–1880) *Family:* Eldest of seven sons; father, a sawmill owner. *Education:* Attended Dickinson. *Employment:* Pennsylvania, 1868–1880. *NAS:* Elected 1876 [entomology]. *Religion:* Reared a Methodist, he became a freethinker in his youth but in 1846 converted to Catholicism. *Evolution:* As early as 1844 he defended the transmutation of species but apparently wrote nothing on the subject in later years. *Publication:* "Enumeration of the Recent Freshwater Mollusca Which Are Common to North America and Europe, with Observations on Species and Their Distribution," *Journal of the Boston Society of Natural History* 4 (1844): 468–484. *Biographies:* Horace L. Haldeman, "A Memoir of Prof. Samuel Steman Haldeman, LL.D.," *Records of the Catholic Historical Society* 9 (1898): 257–271; Francis X. Reuss, "Professor Haldeman as a Catholic," ibid., pp. 271–292. [W. Conner Sorensen]

Hall, James (1811–1898) *Family:* Eldest child in a poor family of five children; father, superintendent of a woolen mill. *Education:* B.S.N., Rensselaer, 1832; M.A., Rennselaer, 1833. *Employment:* New York State Paleontologist, 1843–1898. *NAS:* Elected 1863, mineralogy and geology. *Religion:* Roman Catholic.

Evolution: According to his biographer, "No one was ever able to inveigle Hall into a discussion of the evolution of life." *Biography:* John M. Clarke, *James Hall of Albany: Geologist and Paleontologist, 1811–1898* (Albany, NY: n.p., 1921).

Hayden, Ferdinand Vandiveer (1828–1887) *Family:* Second of four children and eldest surviving son; father, an unstable worker and sometime convict. *Education:* A.B., Oberlin, 1850; M.D., Albany Medical College, 1853. *Employment:* Pennsylvania, 1865–1872; U.S. Geological and Geographical Survey of Territories, 1867–1879; U.S. Geological Survey, 1879–1886. *NAS:* Elected 1873 [geology]. *Religion:* Studied theology at Oberlin in the early 1850s but was never a regular churchgoer; late in life he joined the Mind Cure church, which was similar to Christian Science. *Evolution:* Indirect evidence suggests that he accepted evolution, but he wrote nothing of his views on the subject. *Biographies:* Gerald James Cassidy, "Ferdinand V. Hayden: Federal Entrepreneur of Science," Ph.D. diss., University of Pennsylvania, 1991; Mike Foster, *Strange Genius: The Life of Ferdinand Vandeveer Hayden* (Niwot, CO: Roberts Rinehart, 1994). [Gerald James Cassidy]

Hilgard, Eugene Woldemar (1833–1916) *Family:* Laterborn son in a family of four boys and an unknown number of girls; father, a prominent lawyer and judge. *Education:* Ph.D., Heidelberg, 1853. *Employment:* Mississippi State Geological Survey, 1855–1873; Mississippi, 1866–1873; Michigan, 1873–1875; California, 1874–1905. *NAS:* Elected 1872 [geology and agricultural chemistry]. *Religion:* Joined the Catholic church in 1857 and remained a faithful member. *Evolution:* A theistic evolutionist who stressed the role of deVriesian mutations. *Publication:* "Biographical Memoir of Joseph Le Conte, 1823–1901," *Biographical Memoirs* (National Academy of Sciences) 6 (1909): 147–218, especially pp. 197–198, 210–211. *Biography:* Hans Jenny, *E. W. Hilgard and the Birth of Modern Soil Science* (Pisa: Collana Della Rivista, 1961).

Hitchcock, Edward (1793–1864) *Family:* Third son of a farmer and hatter. *Education:* Attended Yale Theological Seminary. *Employment:* Amherst, 1825–1864. *NAS:* Elected 1863, geology. *Religion:* Ordained Congregational minister. *Evolution:* Rejected. *Publications: Elementary Geology*, new ed. (New York: Ivison, Phinney, 1862), especially pp. 373–393; "The Law of Nature's Constancy Subordinate to the Higher Law of Change," *Bibliotheca Sacra* 20 (1863): 489–561. *Biographies:* Stanley M. Guralnick, "Geology and Religion before Darwin: The Case of Edward Hitchcock, Theologian and Scientist (1793–1864)," *Isis* 63 (1972): 529–543; Rodney Lee Stiling, "The Diminishing Deluge: Noah's Flood in Nineteenth-Century American Thought," Ph.D. diss., University of Wisconsin–Madison, 1991.

Holbrook, John Edwards (1794–1871) *Family:* Elder son from the first marriage of his father, a schoolteacher. *Education:* A.B., Brown, 1815; M.D., Pennsylvania, 1818; additional study in Europe. *Employment:* Professor of anatomy, Medical College of South Carolina, 1824–1854, and physician in Charleston, South Carolina. *NAS:* Elected 1868 [herpetology]. *Religion:* Apparently agnostic; refused to discuss religious topics. *Evolution:* Unknown. *Biography:* Lester D. Stephens, "John Edwards Holbrook and Lewis R. Gibbes: Exemplary Naturalists in the Old South," in *Collection Building in Ichthyology and Herpetology in the Eighteenth, Nineteenth, and Twentieth Centuries,* ed. Theodore W. Pietsch and William D. Anderson, Jr. (Washington, D.C.: American Society of Ichthyologists and Herpetologists, 1997). [Lester D. Stephens]

Hunt, Thomas Sterry (1826–1892) *Family:* Firstborn of at least three children; his merchant father died when Sterry was about twelve, leaving the family impoverished. *Education:* Attended Yale. *Employment:* Geological Survey of Canada, 1846–1872; Laval, 1856–1862; McGill, 1862–1868; MIT, 1872–1878. *NAS:* Elected 1873 [geology and chemistry]. *Religion:* Reared a Congregationalist, he briefly converted to Catholicism as an adult but abandoned formal religion in the 1860s for deism. *Evolution:* As late as 1875 he expressed skepticism about the transmutation of species. *Publication: Chemical and Geological Essays* (Boston: James R. Osgood, 1875). *Biographies:* James Douglas, "Obituary Notice of Thomas Sterry Hunt," *Proceedings of the American Philosophical Society, Memorial Volume* 1 (1900): 63–121; W. H. Brock, "Thomas Sterry Hunt," *Dictionary of Scientific Biography,* 6:564–566. [Suzanne Zeller]

Hyatt, Alpheus (1838–1902) *Family:* Probably first son of a wealthy merchant. *Education:* B.S., Lawrence Scientific School, Harvard, 1862. *Employment:* Harvard Museum of Comparative Zoology, 1865–1902; Essex Institute of Salem, 1867–1869; Peabody Academy of Science, 1869–1870; MIT, 1870–1888. *NAS:* Elected 1875, biology [zoology and paleontology]. *Religion:* After almost converting to Catholicism in the late 1850s, he became, despite a lapse or two into theism, generally skeptical in religious matters. *Evolution:* An early convert to neo-Lamarckian evolution. *Publications:* "On the Parallelism between the Different States of Life in the Individual and Those in the Entire Group of the Molluscous Order Tetrabrachiata," *Memoirs of the Boston Society of Natural History* 1 (1866): 193–209; untitled note, *Proceedings of the Boston Society of Natural History* 14 (1871): 146–148. *Biographies:* Alfred Goldsborough Mayer, "Alpheus Hyatt, 1838–1902," *Popular Science Monthly* 78 (1911): 129–146; Ralph W. Dexter, "The Impact of Evolutionary Theories on the Salem Group of Agassiz Zoologists (Morse, Hyatt, Packard, Putnam)," *Essex Institute Historical Collections* 115 (1979): 144–171; Mary P. Winsor, *Reading the Shape of Nature:*

Comparative Zoology at the Agassiz Museum (Chicago: University of Chicago Press, 1991), p. 41. [Edward Lurie, Lester D. Stephens]

King, Clarence Rivers (1842–1901) *Family:* Only surviving child; his father, a prosperous China trader, died when Clarence was six. *Education:* Ph.B., Sheffield Scientific School, Yale, 1862. *Employment:* California State Geological Survey, 1863–1866; Geological Exploration of the Fortieth Parallel, 1867–1878; U.S. Geological Survey, 1879–1881; independent mining geologist, 1882–1893. *NAS:* Elected 1876, geology and paleontology. *Religion:* A theist who rarely discussed religion as an adult. *Evolution:* Associated evolution with catastrophism. *Publication:* "Catastrophism and Evolution," *American Naturalist* 11 (1877): 449–470. *Biography:* Thurman Wilkins, *Clarence King: A Biography*, rev. ed. (Albuquerque: University of New Mexico Press, 1988).

Kirtland, Jared Potter (1793–1877) *Family:* First of four children; his father, a prosperous stockholder and land agent, left Jared at age ten to be raised by his grandfather, a distinguished physician. *Education:* M.D., Yale, 1815. *Employment:* Cleveland Medical College of Western Reserve College, 1843–1864; medical practice in Cleveland. *NAS:* Elected 1865, zoology. *Religion:* He was "skeptical on all matters of religion." *Evolution:* Unknown. *Biographies:* Agnes Robbins Gehr, "Jared Potter Kirtland," *The Explorer* 2 (1952): 1–33; Herbert Thomas, *The Doctors Jared of Connecticut: Jared Eliot—Jared Potter—Jared Kirtland* (Hamden, CT: Shoe String Press, 1958); Walter B. Hendrickson, *The Arkites and Other Pioneer Natural History Organizations of Cleveland* (Cleveland: Press of Western Reserve University, 1962). [Patsy Gerstner]

LeConte, John Lawrence (1825–1883) *Family:* Only surviving child, whose mother died shortly after his birth; father, an army engineer and naturalist. *Education:* A.B., Mount St. Mary's, 1842; M.D., College of Physicians and Surgeons, New York City, 1846. *Employment:* Financially independent in Philadelphia. *NAS:* Elected 1863, zoology [entomology]. *Religion:* Unknown. *Evolution:* Believed that Providence "presides over and directs the system of evolution." *Publication:* "Address," *American Naturalist* 9 (1875): 481–498. [Lester D. Stephens]

LeConte, Joseph (1823–1901) *Family:* The fifth of six surviving children and the youngest son; father owned a large plantation. *Education:* A.B., Franklin College, Georgia, 1841; M.D., College of Physicians and Surgeons, New York City, 1845; B.S., Lawrence Scientific School, Harvard, 1851. *Employment:* South Carolina, 1857–1862, 1866–1869; California, 1869–1901. *NAS:* Elected 1875, geology and paleontology. *Religion:* An ecumenically minded Presbyterian. *Evolution:* A theistic evolutionist who stressed the "paroxysmal movement of organic evolution." *Publications:* "On Critical Periods in the History of the Earth and Their Relation to Evolution," *American Journal of Science* 114 (1877): 99–114; *Evolution: Its Nature, Its Evidences, and Its Relation to Religious Thought,* 2nd ed. (New York: D. Appleton, 1891). *Biographies: The Autobiography of Joseph LeConte,* ed. William Dallam Armes (New York: D. Appleton, 1903); Lester D. Stephens, *Joseph LeConte: Gentle Prophet of Evolution* (Baton Rouge: Louisiana State University Press, 1982).

Leidy, Joseph (1823–1891) *Family:* The third of four children and the second son; father, a hatter. *Education:* M.D., Pennsylvania, 1844. *Employment:* Pennsylvania, 1853–1891; Swarthmore, 1870–1885. *NAS:* Elected 1863 [anatomy and paleontology]. *Religion:* Abandoned the Lutheranism of his childhood for Unitarianism. *Evolution:* One of the earliest American converts to evolution, which he attributed to multiple causes. *Publication:* "An Address on Evolution and the Pathological Importance of Lower Forms of Life," *Therapeutic Gazette* 10 (1886): 361–358. [Leonard Warren]

Lesley, J. Peter (1819–1903) *Family:* Third of many children and the eldest of eight sons; father, a cabinet-maker. *Education:* A.B., Pennsylvania, 1838; attended Princeton Theological Seminary, 1841–1844; additional study in Germany. *Employment:* Pennsylvania, 1859–1883; American Philosophical Society, 1859–1897; Geological Survey of Pennsylvania, 1873–1887. *NAS:* Elected 1863, mineralogy and geology. *Religion:* Studied for the Presbyterian ministry and pastored a Congregational church but later supported Unitarianism, though he never formally joined that denomination. *Evolution:* After flirting with evolution, he turned his back on it, denouncing it as "the prevalent epidemic scientific superstition of the day." *Publication: Man's Origin and Destiny* (Philadelphia: J. B. Lippincott, 1868). *Biography:* Mary Lesley Ames, ed., *Life and Letters of Peter and Susan Lesley,* 2 vols. (New York: G. P. Putnam's Sons, 1909). [Patsy Gerstner]

Lesquereux, Leo (1806–1889) *Family:* Only son; father, a small manufacturer of watch springs. *Education:* No university training. *Employment:* Columbus, Ohio, working with the bryologist W. S. Sullivant and operating a small jewelry business, 1848–1889; Harvard Museum of Comparative Zoology, 1867–1872. *NAS:* Elected 1864, mineralogy and geology [paleobotany]. *Religion:* Prepared for the Lutheran ministry in his native Switzerland; remained a devout Lutheran throughout his life. *Evolution:* In 1864 he announced his conversion to evolution, but into the 1880s he continued to think that it accounted for the distribution rather than "the first appearance" of species. *Biographies:* Edward Orton, "Leo Lesquereux," *American Geologist* 5 (1890): 284–296; Andrew Denny Rodgers III, *American Botany, 1873–1892: Decades of Transition* (Princeton: Princeton University Press, 1944).

Marsh, Othniel Charles (1831–1899) *Family:* Second child and first son in a family of three surviving children, plus six younger step-brothers and step-sisters; father, a farmer. *Education:* A.B., Yale, 1860; attended Sheffield Scientific School, 1861–62; additional study in Germany. *Employment:* Yale, 1866–1899; U.S. Geological Survey, 1882–1899. *NAS:* Elected 1874, biology [paleontology]. *Religion:* Unknown. *Evolution:* An ardent evolutionist who believed the process was driven by an inherently progressive force. *Publications:* "Fossil Horses in America," *American Naturalist* 8 (1874): 288–294; *Odontornithes: A Monograph on the Extinct Toothed Birds of North America, United States Geological Exploration 40th Parallel* (1880), vol. 7; "The Dinosaurs of North America," *Report of the U.S. Geological Survey* 16, part 1 (1896): 133–414. *Biography:* Charles Schuchert and Clara Mae LeVene, *O. C. Marsh: Pioneer in Paleontology* (New Haven: Yale University Press, 1940).

Meek, Fielding Bradford (1817–1876) *Family:* Laterborn in a family of four children; father, a lawyer. *Education:* No collegiate training. *Employment:* Smithsonian Institution, 1858–1876. *NAS:* Elected 1869 [paleontology]. *Religion:* Unknown. *Evolution:* An extremely reticent man, he seems to have left no record of his views on the subject, though some indirect evidence suggests he may have accepted evolution. [Clifford M. Nelson]

Minot, Charles Sedgwick (1852–1914) *Family:* Second son in a large family; raised in wealth on a country estate. *Education:* B.S., MIT, 1872; study in Europe, 1873–1876; Sc.D. Harvard, 1878. *Employment:* Harvard, 1880–1914. *NAS:* Elected 1897, biology [anatomy]. *Religion:* A Christian of some kind. *Evolution:* Accepted evolution by 1870; in the 1880s disagreed with Weismann on the inheritance of acquired characters. *Publications:* "On the Conditions to Be Filled by a Theory of Life," *Proceedings of the American Association for the Advancement of Science* 28 (1880): 411–415; "Charles Robert Darwin," *Boston Medical and Surgical Journal* 106 (1882): 402–403. *Biography:* Frederic T. Lewis, "Charles Sedgwick Minot," *Anatomical Record* 10 (1916): 133–164. [Philip J. Pauly]

Mitchell, Silas Weir (1829–1914) *Family:* The third of nine children; father, a physician. *Education:* M.D., Jefferson Medical College, 1850; additional study in Europe. *Employment:* medical practice in Philadelphia. *NAS:* Elected 1865, anthropology [neurology]. *Religion:* Episcopalian. *Evolution:* Unknown. *Biographies:* Anna Robeson Burr, *Weir Mitchell: His Life and Letters* (New York: Duffield, 1929); Ernest Earnest, *S. Weir Mitchell: Novelist and Physician* (Philadelphia: University of Pennsylvania Press, 1950). [Bonnie Blustein]

Morgan, Lewis Henry (1818–1881) *Family:* Ninth of thirteen children; father, a prosperous farmer and state senator. *Education:* B.A., Union, 1840. *Employ-*

ment: Legal practice in Rochester, New York. *NAS:* Elected 1875 [anthropology]. *Religion:* Attended a Presbyterian church and declared "my heart is with the Christian religion." *Evolution:* Some evidence suggests that he accepted biological evolution by the early 1870s, but his exact views are frustratingly difficult to determine. *Biographies:* J. H. McIlvaine, *The Life and Works of Lewis H. Morgan* (n.p., n.d.); Carl Resek, *Lewis Henry Morgan: American Scholar* (Chicago: University of Chicago Press, 1960); Adam Kuper, "The Development of Lewis Henry Morgan's Evolutionism," *Journal of the History of the Behavioral Sciences* 21 (1985): 3–22; Thomas R. Trautmann, *Lewis Henry Morgan and the Invention of Kinship* (Berkeley and Los Angeles: University of California Press, 1987).

Morse, Edward Sylvester (1838–1925) *Family:* Third child and third son in a family of six children; father, a successful fur merchant. *Education:* Attended Lawrence Scientific School, Harvard, 1859–1862. *Employment:* Harvard Museum of Comparative Zoology, 1862–1866; Peabody Academy of Science, 1867–1870; Bowdoin, 1871–1874; Tokyo, 1877–1880; Peabody Museum, 1880–1925. *NAS:* Elected 1876, biology [zoology and ethnology]. *Religion:* Son of a Baptist deacon, he became a freethinker in the late 1850s. *Evolution:* After accepting evolution in 1873, he became one of its most fervent apostles in America, arguing for the direct modifying influence of the environment. *Publications:* "Address," *Proceedings of the American Association for the Advancement of Science* 25 (1876): 136–176; "Address," ibid. 36 (1887): 1–43. Both addresses also appeared under the title "What American Zoologists Have Done for Evolution." *Biographies:* Dorothy G. Wyman, *Edward Sylvester Morse: A Biography* (Cambridge: Harvard University Press, 1942); Mary P. Winsor, *Reading the Shape of Nature: Comparative Zoology at the Agassiz Museum* (Chicago: University of Chicago Press, 1991), p. 41.

Newberry, John Strong (1822–1892) *Family:* Youngest of nine children; father, a wealthy landowner and entrepreneur. *Education:* A.B., Western Reserve College, 1846; M.D., Cleveland Medical School, 1848; additional study in Paris. *Employment:* U.S. Sanitary Commission, 1861–1865; Columbia University School of Mines, 1866–1892. *NAS:* Elected 1863, mineralogy and geology. *Religion:* Very religious, but denominational affiliation unknown. *Evolution:* Rejected evolution in his only known public statement on the subject, made in 1867. *Publication:* "Address," *Proceedings of the American Association for the Advancement of Science* 16 (1867): 1–15. [Michael C. Hansen]

Osborn, Henry Fairfield (1857–1935) *Family:* The second of four children and the first of three sons; father, a wealthy railroad magnate. *Education:* A.B., Princeton, 1877; additional training in New York City, 1878–1879, and Europe, 1879–1880; Sc.D., Princeton, 1881. *Employment:* Princeton, 1881–1890; Columbia, 1891–1910; American Museum of Natural History, 1891–1935. *NAS:* Elected 1900, geology, paleontology, and biology. *Religion:* Attended a Presbyterian church in Princeton, Episcopal services in New York City. *Evolution:* A theistic evolutionist, he espoused neo-Lamarckian views until the 1890s. *Publications:* "The Paleontological Evidence for the Transmission of Acquired Characters," *American Naturalist* 23 (1889): 559–566; "The Hereditary Mechanism and the Search for the Unknown Factors of Evolution," ibid. 29 (1895): 418–439; "The Limits of Organic Selection," ibid. 31 (1897): 944–951; "Ontogenic and Phylogenic Variation," *Science,* n.s., 4 (1896): 786–789; "The Nine Principles of Evolution Revealed by Paleontology," *American Naturalist* 66 (1932): 52–60. *Biography:* Ronald Rainger, *An Agenda for Antiquity: Henry Fairfield Osborn and Vertebrate Paleontology at the American Museum of Natural History, 1890–1935* (Tuscaloosa: University of Alabama Press, 1991). [Ronald Rainger]

Packard, Alpheus Spring, Jr. (1839–1905) *Family:* Youngest of five children, including three brothers; father, a minister and college professor. *Education:* A.B., Bowdoin, 1861; A.M., Bowdoin, 1862; S.B., Lawrence Scientific School, Harvard, 1864; M.D., Maine Medical School, Bowdoin, 1864. *Employment:* Various short-term appointments, 1864–1878; Brown, 1878–1905. *NAS:* Elected 1872, geology, paleontology, and biology [entomology]. *Religion:* The son of a Congregational minister, he remained a theist as an adult; denominational affiliation, if any, is unknown. *Evolution:* One of the founders of the so-called neo-Lamarckian school. *Publications:* (with F. W. Putnam) *The Mammoth Cave and Its Inhabitants; or, Descriptions of the Fishes, Insects, and Crustaceans Found in the Cave* (Salem, MA: Naturalists' Agency, 1872); "The Law of Evolution," *Independent* 32 (February 5, 1880): 10; "A Half-Century of Evolution, with Special Reference to the Effects of Geological Changes on Animal Life," *Proceedings of the American Association for the Advancement of Science* 47 (1898): 311–356; *Lamarck, the Founder of Evolution: His Life and Work* (New York: Longmans, Green, 1901).

Pourtalès, Louis François de (1824–1880) *Family:* Son of a wealthy European count; birth order unknown. *Education:* Trained as an engineer in Switzerland. *Employment:* U.S. Coast Survey, 1848–1873; Harvard Museum of Comparative Zoology, 1870–1880. *NAS:* Elected 1873 [oceanography]. *Religion:* Unknown. *Evolution:* Unknown. According to his friend Alexander Agassiz, "He never entered into a single scientific controversy." *Publication:* "Galápagos," in *The American Cyclopaedia: A Popular Dictionary of General Knowledge,* ed. George Ripley and Charles A. Dana (New York: D. Appleton, 1874), 7:556–557.

Powell, John Wesley (1834–1902) *Family:* The fourth of nine children; father, a Methodist minister. *Education:* Attended Illinois Institute (now Wheaton College), Illinois College, and Oberlin. *Employment:* Illinois Wesleyan, 1865; Illinois State Normal University, 1866–1872; Geographical and Geological Survey of the Rocky Mountain Region, 1870–1879; Smithsonian Institution, 1879–1902; U.S. Geological Survey, 1881–1894. *NAS:* Elected 1880, anthropology [geology]. *Religion:* Raised a Methodist, as an adult he became an atheist. *Evolution:* He accepted a vaguely Lamarckian form of biological evolution, even in explaining the origin of the human body, but denied that the "laws of biotic evolution" applied to human society. *Publications:* "Darwin's Contributions to Philosophy," *Proceedings of the Biological Society of Washington* 1 (1880–1882): 60–70; "Human Evolution," *Transactions of the Anthropological Society of Washington* 2 (1883): 176–208; "The Three Methods of Evolution," *Bulletin of the Philosophical Society of Washington* 6 (1884): xxvii–lii; "From Barbarism to Civilization," *American Anthropologist* 1 (1888): 97–123; "Competition as a Factor in Human Evolution," ibid., pp. 297–323. *Biographies:* William Culp Darrah, *Powell of the Colorado* (Princeton: Princeton University Press, 1951); Wallace Stegner, *Beyond the Hundredth Meridian: John Wesley Powell and the Second Opening of the West* (Boston: Houghton Mifflin, 1954); John Joseph Zernel, "John Wesley Powell: Science and Reform in a Positive Context," Ph.D. diss., Oregon State University, 1983. [John J. Zernel]

Pumpelly, Raphael (1837–1923) *Family:* Youngest of four children and second son; father, a successful merchant and businessman. *Education:* Studied at the Bergakademie, Freiburg, 1856–1860. *Employment:* Geologist and mining engineer in Japan and China, 1861–1865; various explorations of the Lake Superior region, 1866–1877; U.S. Geological Survey, 1879–1881, 1884–1890; Northern Transcontinental Survey, 1881–1884; consulting and traveling, 1890–1923. *NAS:* Elected 1872, geology and paleontology. *Religion:* Rejected the Presbyterianism of his youth for religious skepticism and occasional mysticism. *Evolution:* Unknown. *Publication: My Reminiscences* (New York: Henry Holt, 1918). *Biography:* Peggy Champlin, *Raphael Pumpelly: Gentlemen Geologist of the Gilded Age* (Tuscaloosa: University of Alabama Press, 1994).

Putnam, Frederic Ward (1839–1915) *Family:* Youngest child and third son among four surviving children; father, a horticulturalist and postmaster. *Education:* S.B., Harvard, 1862. *Employment:* Essex Institute of Salem, 1856–1870; Peabody Academy of Science, 1869–1873; Peabody Museum of American Archaeology and Ethnology, Harvard, 1875–1915; American Museum of Natural History, 1894–1903; California, 1903–1909. *NAS:* Elected 1885, biology and anthropology. *Religion:* Unitarian. *Evolution:* A reluctant evolutionist. *Publication:* (with A. S. Packard, Jr.) *The Mammoth Cave and Its Inhabitants; or, Descriptions of the Fishes, Insects and Crustaceans Found in the Cave* (Salem, MA: Naturalists' Agency, 1872). *Biography:* Ralph W. Dexter, "The Impact of Evolutionary

Theories on the Salem Group of Agassiz Zoologists (Morse, Hyatt, Packard, Putnam)," *Essex Institute Historical Collections* 115 (1979): 144–171. [Terry A. Barnhart]

Rogers, William Barton (1804–1882) *Family:* Second of four surviving children, all boys; father, a physician and college professor. *Education:* Attended College of William and Mary. *Employment:* MIT, 1862–1881. *NAS:* Elected 1863 [geology]. *Religion:* If anything, a Presbyterian, though he had close ties to Unitarians. *Evolution:* Even before 1859 he believed in "the gradual modification of species through external conditions," especially "violent and sudden physical changes." *Publications:* Review of *On the Origin of Species,* by Charles Darwin, *Boston Courier,* March 5, 1860; *Proceedings of the Boston Society of Natural History* 7 (1859–1861): 231–274. *Biography:* Emma Rogers, *Life and Letters of William Barton Rogers,* 2 vols. (Boston: Houghton Mifflin, 1896). [Loretta H. Mannix]

Sargent, Charles Sprague (1841–1927) *Family:* Youngest of four children and the second son; father, a merchant and banker. *Education:* A.B., Harvard, 1862. *Employment:* Harvard, 1872–1927. *NAS:* Elected 1895, biology [botany]. *Religion:* Unknown, though married to a devout Episcopalian. *Evolution:* Unknown. *Biography:* S. B. Sutton, *Charles Sprague Sargent and the Arnold Arboretum* (Cambridge: Harvard University Press, 1970).

Scudder, Samuel Hubbard (1837–1911) *Family:* Laterborn son in a family of seven children; father, a well-known hardware merchant. *Education:* A.B., Williams, 1857; A.M., Williams, 1860; S.B., Harvard, 1862. *Employment:* Boston Society of Natural History, 1862–1887; U.S. Geological Survey, 1886–1892. *NAS:* Elected 1877, biology [entomology]. *Religion:* Reared a Congregationalist; adult affiliation unknown. *Evolution:* Became a theistic evolutionist in the early 1870s. *Publications:* "Is Mimicry Advantageous?" *Nature* 3 (1870): 147; "Address," *Proceedings of the American Association for the Advancement of Science* 29 (1880): 609–615; *The Butterflies of the Eastern United States and Canada, with Special Reference to New England,* 3 vols. (Cambridge, MA: By the Author, 1889). *Biography:* W. Conner Sorensen, *Brethren of the Net: American Entomology, 1840–1880* (Tuscaloosa: University of Alabama Press, 1995), pp. 199, 204. [W. Conner Sorensen]

Silliman, Benjamin, Sr. (1779–1864) *Family:* Younger son of a lawyer and Revolutionary War general, who died when Benjamin was eleven years old. *Education:* A.B., Yale, 1796; additional study in Philadelphia, Princeton, London, and Edinburgh. *Employment:* Yale, 1799–1853. *NAS:* Elected 1863, mineralogy and geology. *Religion:* Congregationalist. *Evolution:* An antievolutionist who probably never read the *Origin of Species. Biographies:* John F. Fulton and Elizabeth H. Thomson, *Benjamin Silliman, 1779–1864: Pathfinder in American Science* (New York: Henry Schuman, 1947); Chandos Michael Brown, *Benjamin Silliman: A Life in the Young Republic* (Princeton: Princeton University Press, 1989). [Chandos Michael Brown]

Smith, Sidney Irving (1843–1926) *Family:* One of at least three children born to an old New England family. *Education:* Ph.B., Sheffield Scientific School, Yale, 1867. *Employment:* Yale, 1867–1906. *NAS:* Elected 1884, biology [comparative anatomy]. *Religion:* Unknown. *Evolution:* Described as an "eager" evolutionist.

Stimpson, William (1832–1872) *Family:* Son of a prosperous merchant and inventor; birth order uncertain (he was either the eldest or second eldest of four children). *Education:* Studied with Louis Agassiz at Harvard. *Employment:* Smithsonian Institution, 1856–1865; Chicago Academy of Science, 1865–1871. *NAS:* Elected 1868 [conchology]. *Religion:* Episcopalian. *Evolution:* Accepted evolution but left few traces of his views. *Publication:* Review of "The Kjökkenmöddings: Recent Geologico-Archaeological Researches in Denmark," by John Lubbock, *American Journal of Science,* 83 (1862): 297–298. *Biography:* Ronald S. Vasile, *A Fine View from the Shore: The Life of William Stimpson,* forthcoming. [Ronald S. Vasile]

Sullivant, William Starling (1803–1873) *Family:* Eldest of three sons of a wealthy surveyor and landowner. *Education:* A.B., Yale, 1823. *Employment:* Financially independent in Columbus, Ohio. *NAS:* Elected 1872 [bryology]. *Religion:* Probably a theist, though not a churchman. *Evolution:* Unknown; never mentioned evolution in his correspondence. *Biography:* Andrew Denny Rodgers III, *"Noble Fellow": William Starling Sullivant* (New York: G. P. Putnam's Sons, 1940).

Torrey, John (1796–1873) *Family:* Second son in a family of nine children whose elder brother died before adulthood; father, a merchant and fiscal agent for a state prison. *Education:* M.D., College of Physicians and Surgeons, New York City, 1818. *Employment:* U.S. Assay Office in New York City, 1853–1873. *NAS:* Elected 1863, chemistry and botany. *Religion:* Lifelong Presbyterian. *Evolution:* Delayed reading the *Origin of Species* and never expressed his views. *Biographies:* Andrew Denny Rodgers III, *John Torrey: A Story of North American Botany* (Princeton: Princeton University Press, 1942); Christine Chapman Robbins, "John Torrey (1796–1873): His Life and Times," *Bulletin of the Torrey Botanical Club* 95 (1968): 515–645.

Tuckerman, Edward (1817–1886) *Family:* Eldest son; father, an apparently successful merchant. *Education:* A.B., Union, 1837; LL.B., Harvard Law School; M.A., Union, 1843; A.B., Harvard, 1847; completed program at Harvard Divinity School, 1852. *Employment:* Amherst, 1855–1886. *NAS:* Elected 1868 [botany]. *Religion:* Lifelong Episcopalian. *Evolution:* Unknown, though his early reaction was negative. *Writings:* Edward Tuckerman to Asa Gray, March 3, 1860, and April 26, 1877, Historic Letters File, Library of the Gray Herbarium, Harvard University, Cambridge. *Biography:* Anna M.M. Reid, "Edward Tuckerman (1817–1886), Pioneer American Lichenologist: The Early Years" *Mycotaxon* 16 (July/September, 1986): 3–16.

Verrill, Addison Emery (1839–1926) *Family:* Second son in a family of eight children; father, a merchant and carpenter. *Education:* B.S., Lawrence Scientific School, Harvard, 1862. *Employment:* Harvard Museum of Comparative Zoology, 1861–1864; Yale, 1864–1907. *NAS:* Elected 1872, biology [zoology]. *Religion:* A theist who attended the Yale College chapel with his family but never joined. *Evolution:* Accepted evolution; a close friend of Alpheus Hyatt. *Publications:* "Parental Instinct as a Factor in the Evolution of Species," *Science* 1 (1883): 303–304; "Nocturnal Protective Coloration in Mammals, Birds, Fishes, Insects, Etc., as Developed by Natural Selection," *American Journal of Science* 153 (1897): 132–134; "Remarkable Development of Starfishes on the Northwest American Coast; Hybridism; Multiplicity of Rays; Teratology; Problems in Evolution; Geographical Distribution," *American Naturalist* 43 (1909): 542–555. *Biography:* George E. Verrill, *The Ancestry, Life, and Work of Addison E. Verrill of Yale University* (Santa Barbara: Pacific Coast Publishing, 1958). [Lester D. Stephens]

Walcott, Charles Doolittle (1850–1927) *Family:* Youngest of four children whose father, a businessman, died when Charles was two years old. *Education:* Studied privately with James Hall. *Employment:* U.S. Geological Survey, 1879–1907; Smithsonian Institution, 1897–98, 1907–1927. *NAS:* Elected 1896, geology, paleontology, biology. *Religion:* Devout Presbyterian. *Evolution:* A theistic evolutionist who stressed the role of environmental changes. *Publications:* "Geologic Time: As Indicated by the Sedimentary Rocks of North America," *American Geologist* 12 (1893): 343–368; "Evolution of Early Paleozoic Faunas in Relation to Their Environment," *Journal of Geology* 17 (1909): 193–202. *Biography:* Stephen Jay Gould, *Wonderful Life: The Burgess Shale and the Nature of History* (New York: W. W. Norton, 1989), pp. 240–262.

Watson, Sereno (1826–1892) *Family:* Ninth child and eighth son in a family of thirteen children; father, a merchant and farmer. *Education:* A.B., Yale, 1847; additional study at Sheffield Scientific School, Yale, 1866–67. *Employment:* Geological Exploration of the Fortieth Parallel, 1867–68; Harvard, 1870–1892, the first three years unofficially. *NAS:* Elected 1889, biology [botany]. *Religion:* Faithful Congregationalist. *Evolution:* Unknown. [A. Hunter Dupree]

Welch, William Henry (1850–1934) *Family:* Second child and only son, whose mother died when he was about six months old; raised as an only child by a grandmother; father, a physician. *Education:* A.B., Yale, 1870; M.D., College of Physicians and Surgeons, New York City, 1875; additional study in Europe. *Employment:* Bellevue Hospital Medical College, 1878–1884; Johns Hopkins, 1884–1931. *NAS:* Elected 1895, biology and anthropology [pathology and bacteriology]. *Religion:* An active Congregationalist in his youth, he quit attending church in his mid-twenties and apparently became an agnostic. *Evolution:* Embraced evolution in the mid-1870s; regarded "variation and natural selection" as an important, but insufficient, factor in explaining evolution. *Publication:* "Adaptation in Pathological Processes," in *Papers and Addresses by William Henry Welch,* 3 vols. (Baltimore: Johns Hopkins University Press, 1920), 1:370–394. *Biographies:* Simon Flexner and James Thomas Flexner, *William Henry Welch and the Heroic Age of American Medicine* (New York: Viking Press, 1941); Donald Fleming, *William H. Welch and the Rise of Modern Medicine* (Boston: Little, Brown, 1954).

White, Charles Abiathar (1826–1910) *Family:* Second son; father's occupation unknown. *Education:* M.D., Rush Medical College, 1864. *Employment:* Iowa, 1867–1873; Bowdoin, 1873–1875; Geographical and Geological Survey of the Rocky Mountain Region, 1875; U.S. Geological and Geographical Survey of the Territories, 1876–1879; U.S. National Museum, Smithsonian Institution, 1879–1882; U.S. Geological Survey, 1883–1892. *NAS:* Elected 1889, geology, paleontology, and biology. *Religion:* Unknown. *Evolution:* Favored deVries's mutation theory over natural selection, because the fossil evidence for both plants and animals seemed "to be quite inconsistent with the theory of their origin by the slow process of natural selection." *Publications:* "The Relation of Biology to Geological Investigations," *Annual Report of the Smithsonian Institution,* 1892, pp. 245–368; "The Mutation Theory of Professor De Vries," ibid., 1901, pp. 631–640; "The Saltatory Origin of Species," *Bulletin of the Torrey Botanical Club* 29 (1902): 511–522. [Daniel Goldstein]

Whitman, Charles Otis (1842–1910) *Family:* Eldest of four children; father, a carriage maker. *Education:* B.A., Bowdoin, 1868; Ph.D., Leipzig, 1878. *Employment:* Imperial University of Japan, 1879–1881; Harvard Museum of Comparative Zoology, 1882–1886; Allis Lake Laboratory, Milwaukee, 1886–1889; Marine Biological Laboratory, Woods Hole, Massachusetts, 1888–1908; Clark, 1889–1892; Chicago, 1892–1910. *NAS:* Elected 1895, biology [zoology]. *Religion:*

Rejected the Adventism of his father for agnosticism. *Evolution:* A leading advocate of orthogenesis, assisted by natural selection. *Publications:* "The Problem of the Origin of Species," in *Congress of Arts and Science: Universal Exposition, St. Louis, 1904,* ed. Howard J. Rodgers, 8 vols. (Boston: Houghton Mifflin, 1905–1907), 5:41–58; "The Origin of Species," *Bulletin of the Wisconsin Natural History Society* 5 (1907): 6–14. *Biography:* Philip J. Pauly, "From Adventism to Biology: The Development of Charles Otis Whitman," *Perspectives in Biology and Medicine* 37 (1994): 395–408.

Whitney, Josiah Dwight (1819–1896) *Family:* Eldest of eight children; father, a prosperous banker. *Education:* A.B., Yale, 1839; additional study in Europe. *Employment:* California State Geological Survey, 1860–1874; Harvard, 1865–1896. *NAS:* Elected 1863, mineralogy and geology. *Religion:* Unitarian with skeptical leanings. *Evolution:* Defended Darwin as early as 1862 but left little record of his views thereafter. *Publication: Lecture on Geology: Delivered before the Legislature of California, at San Francisco, Thursday Evening, Feb. 27, 1862* (San Francisco: Benj. P. Avery, 1862). *Biography:* Edwin Tenney Brewster, *Life and Letters of Josiah Dwight Whitney* (Boston: Houghton Mifflin, 1909). [Stephen G. Alter]

Wilson, Edmund Beecher (1856–1939) *Family:* Fourth child and third son in a family of five children; father, a lawyer and judge. *Education:* Ph.B., Sheffield Scientific School, Yale, 1878; Ph.D., Johns Hopkins, 1881. *Employment:* Williams, 1883–84; MIT, 1884–85; Bryn Mawr, 1885–1891; Columbia, 1891–1928. *NAS:* Elected 1899, biology [embryology and cytology]. *Religion:* Unknown. *Evolution:* Accepted evolution as a "fact" but found natural selection inadequate to explain it. *Publications: The Cell in Development and Inheritance* (New York: Macmillan, 1896); *Biology* (New York: Columbia University Press, 1908); Introduction to *The Origin of Species by Means of Natural Selection* (New York: Macmillan, 1927). *Biographies:* H. J. Muller, "Edmund B. Wilson—An Appreciation," *American Naturalist* 77 (1943): 142–172; Alice Baxter Levine, "Edmund Beecher Wilson and the Problem of Development: From the Germ Layer Theory to the Chromosome Theory of Inheritance," Ph.D. diss., Yale University, 1974; Garland E. Allen, "Edmund Beecher Wilson," *Dictionary of Scientific Biography,* 14:423–436; Jane Maienschien, *Transforming Traditions in American Biology, 1880–1915* (Baltimore: Johns Hopkins University Press, 1991). [Garland E. Allen]

Wood, Horatio C (1841–1920) *Family:* Son of a successful businessman; birth order unknown. *Education:* M.D., Pennsylvania, 1862. *Employment:* Pennsylvania, 1866–1906. *NAS:* Elected 1879, biology [pharmacology]. *Religion:* Lifelong Methodist. *Evolution:* Unknown. *Biographies:* George B. Roth, "An Early American Pharmacologist: Horatio C Wood (1841–1920)," *Isis* 30 (1939): 38–45; Glenn Sonnedecker, "Horatio C Wood," *Dictionary of Scientific Biography,* 14:495–497. [John P. Swann]

Woodward, Joseph Janvier (1833–1884) *Family:* Eldest of three children; father's occupation unknown. *Education:* A.B., Central High School of Philadelphia, 1850; M.D., Pennsylvania, 1853; M.A., Central High School of Philadelphia, 1855. *Employment:* U.S. Surgeon General's Office. *NAS:* Elected 1873 [microscopy and histology]. *Religion:* Unknown. *Evolution:* Accepted evolution but believed in the supernatural creation of life. *Publication:* "Modern Philosophical Conceptions of Life," *Bulletin of the Philosophical Society of Washington* 5 (1883): 49–84. [Mary C. Gillett]

Worthen, Amos Henry (1813–1888) *Family:* Eleventh of twelve children; father, a farmer. *Education:* No collegiate training. *Employment:* Geological Survey of Illinois, 1858–1888. *NAS:* Elected 1872 [geology]. *Religion:* Unknown. *Evolution:* Indirect evidence suggests that he may have accepted evolution.

Wyman, Jeffries (1814–1874) *Family:* Third son in a family of four boys and one girl; father, a physician. *Education:* A.B., Harvard, 1833; M.D., Harvard Medical School, 1837; additional study in Europe. *Employment:* Harvard, 1847–1874. *NAS:* Elected 1863, anatomy and physiology. *Religion:* Devout Unitarian. *Evolution:* A theistic evolutionist, he was influenced more by Owen than by Darwin. *Publication:* Review of *Monograph of the Aye-Aye,* by Richard Owen, *American Journal of Science* 86 (1863): 294–299. *Biographies:* Burt G. Wilder, "Jeffries Wyman, Anatomist (1814–1874)," in *Leading American Men of Science,* ed. David Starr Jordan (New York: Henry Holt, 1910), pp. 171–209; A. Hunter Dupree, "Some Letters from Charles Darwin to Jeffries Wyman," *Isis* 42 (1951): 104–110; A. Hunter Dupree, "Jeffries Wyman's Views on Evolution," ibid. 44 (1953): 243–246; Toby A. Appel, "Jeffries Wyman, Philosophical Anatomy, and the Scientific Reception of Darwin in America," *Journal of the History of Biology* 21 (1988): 69–94.

NOTES

Introduction

I am especially grateful to Stephen C. Meyer for sharing sources about the intelligent-design movement and to Michael G. Fisher, Jack W. Haas, Jr., and Paul A. Nelson for their critical reading of an earlier draft of this chapter. I am also indebted to the following for sharing information and documents related to the events discussed: David Berlinski, Paul Boyer, Doug Cumming, Jack D. Ellis, Mona Frederick, Carolyn Hackler, Dennis Hamm, S. J., David Lee, Charles E. Rosenberg, Carol Scott, Eugenie C. Scott, Rodney L. Stiling, and Sean Sullivan.

1. See Jon H. Roberts, *Darwinism and the Divine in America: Protestant Intellectuals and Organic Evolution, 1859–1900* (Madison: University of Wisconsin Press, 1988); and Ronald L. Numbers, *The Creationists* (New York: Alfred A. Knopf, 1992).

2. R. Scott Appleby, "Exposing Darwin's 'Hidden Agenda': Roman Catholic Responses to Evolution, 1875–1925," in *Darwin's Reception: The Role of Place, Race, Religion, and Gender*, ed. Ronald L. Numbers and John Stenhouse (New York: Cambridge University Press, 1999); Marc Swetlitz, "American Jewish Responses to Darwin and Evolutionary Theory, 1860–1890," ibid.

3. Documentation for this and the following paragraph can be found in Numbers, *The Creationists.*

4. Gregg A. Mitman and Ronald L. Numbers, "Evolutionary Theory," in *Encyclopedia of the United States in the Twentieth Century*, ed. Stanley I. Kutler, 4 vols. (New York: Charles Scribner's Sons, 1996), 2:859–876; William B. Provine, "Progress in Evolution and Meaning in Life," in *Julian Huxley: Biologist and Statesman of Science*, ed. C. Kenneth Waters and Albert Van Helden (Houston: Rice University Press, 1992), pp. 165–180, quotation on p. 179.

5. Documentation for this and the next seven paragraphs can be found in Numbers, *The Creationists.*

6. The best source on this subject is Edward J. Larson, *Trial and Error: The American Controversy over Creation and Evolution*, updated ed. (New York: Oxford University Press, 1989).

7. Frank J. Sonleitner, "Creationists Embarrassed in Oklahoma," *Creation/Evolution* (Spring 1981): 22–27.

8. Eugenie C. Scott, "In the Trenches," *NCSE Reports* 13 (Summer 1993): 6; Eugenie C. Scott, "Creationist Cases Blooming," ibid. 12 (Summer 1992): 1, 3, 5. NCSE stands for National Center for Science Education.

9. Eugenie C. Scott, "Big Bang Glue-on in Kentucky," *NCSE Reports* 16 (Summer 1996): 1, 9; Karen Schmidt, "The Battle of the Books," *Science* 273 (1996): 421.

10. "Alabama School Board Votes to Put Evolution Message in Biology Texts," Associated Press news release, November 10, 1995. See also Eugenie C. Scott, "State of Alabama Distorts Science, Evolution," *NCSE Reports* 15 (Winter 1995): 10–11. Other states debated the wisdom of following Alabama's lead.

11. Jill Nelson, "Creationism: The Debate Is Still Evolving," *USA Weekend*, April 18–20, 1997, p. 12; Molleen Matsumura, "Textbook Evolution Disclaimer in Fairfax County, VA," *NCSE Reports* 16 (Fall 1996): 16; Molleen Matsumura, "Georgia: Creationism Pushed at State and Local Levels," ibid. 15 (Winter 1995): 8–9; Molleen Matsumura and Andrew J. Petto, "New Anti-Evolution Strategy Rejected by New Hampshire Legislature," ibid. 16 (Spring 1996): 20; Molleen Matsumura, "New Mexico: State Legislature Joins the Fray," *Reports of the National Center for Science Education* 17 (January/February 1997): 4; "Update," ibid., pp. 5–6; Molleen Matsumura, "Tennessee Upset: 'Monkey Bill' Law Defeated," *NCSE Reports* 15 (Winter 1995): 6–7; Duren Cheek, "Bill May Evolve into Law," *Nashville Tennessean*, February 27, 1996, pp. 1A–2A, quotation on p. 1A (rocket); Eugenie C. Scott, "Close Ohio Vote Scuttles 'Evidence against Evolution' Bill," *NCSE Reports* 16 (Spring 1996): 18 (vampire).

12. Numbers, *The Creationists*, p. 300 (Reagan); "Pat Buchanan Takes on Darwin," *NCSE Reports* 15 (Winter 1995): 3–4 (apes); Michael D. Lemonick, "Dumping on Darwin," *Time*, March 18, 1996, p. 81 (Godless evolution); Molleen Matsumura, "Evolution in an Election Year," *NCSE Reports* 14 (Fall 1994): 3, 10.

13. Eugenie C. Scott, "Gallup Reports High Level of Belief in Creationism," *NCSE Reports* 13 (Fall 1993): 9; John Cole, "Gallup Poll Again Shows Confusion," ibid. 16 (Spring 1996): 9; "God Is Alive," *Maclean's*, April 12, 1993, p. 53.

14. "Creationism in NZ 'Unlikely,'" *New Zealand Herald*, July 3, 1986, p. 14 (quoting Gould); Richard C. Lewontin, Introduction to *Scientists Confront Creationism*, ed. Laurie R. Godfrey (New York: W. W. Norton, 1983), p. xxv; Numbers, *The Creationists*, pp. 323–335. See also Ronald L. Numbers and John Stenhouse, "Antievolutionism in the Antipodes: From Protesting Evolution to Promoting Creationism in New Zealand," *British Journal for the History of Science*, in press; and Ronald L. Numbers, "Creationists and Their Critics in Australia: An Autonomous Culture or 'the USA with Kangaroos'?" unpublished paper presented in a session entitled "The Cultures of Creationism" at the annual meeting of the American Anthropological Association, San Francisco, November 24, 1996.

15. Taner Edis, "Islamic Creationism in Turkey," *Creation/Evolution* 14 (Summer 1994): 3–12.

16. Richard Dawkins, *The Blind Watchmaker* (New York: W. W. Norton, 1986), pp. 5–6, 316; Richard Dawkins, Review of *Blueprints: Solving the Mystery of Evolution*, by Maitland A. Edey and Donald C. Johanson, *New York Times*, April 9, 1989,

section 7, p. 34 (ignorant); Roger Downey, "Darwin's Watchdog," *Eastsideweek*, December 11, 1996. In this free newspaper distributed in the Seattle area, Downey describes Dawkins as "a point man for evolution." See also Richard Dawkins, *Climbing Mount Improbable* (New York: W. W. Norton, 1996).

17. Daniel C. Dennett, *Darwin's Dangerous Idea: Evolution and the Meaning of Life* (New York: Simon and Schuster, 1995), pp. 515–516, 519–521. See also Daniel C. Dennett, "Appraising Grace: What Evolutionary Good is God?" *The Sciences* 37 (January/February 1997): 39–44.

18. Stephen Jay Gould, "Nonoverlapping Magisteria," *Natural History* 106 (March 1997): 16–22, 60–62; Stephen Jay Gould, "Impeaching a Self-Appointed Judge," *Scientific American*, July 1992, pp. 118–120; Numbers, *The Creationists*, p. 281 (regarding Wise).

19. Numbers, *The Creationists*, pp. 178–179.

20. Howard J. Van Till, *The Fourth Day: What the Bible and the Heavens Are Telling Us about the Creation* (Grand Rapids, MI: William B. Eerdmans, 1986), pp. 223, 252–253; Del Ratzsch, *The Battle of Beginnings: Why Neither Side Is Winning the Creation-Evolution Debate* (Downers Grove, IL: InterVarsity Press, 1996), p. 180 (theistic evolution confused); Joel Belz, "Witness for the Prosecution," *World*, November 30–December 7, 1996, p. 18 (backwater positions). Like semantically sensitive evangelicals who used to say that they were promoting "progressive creation" rather than "theistic evolution," Van Till prefers to talk about advocating "the creationomic perspective," not "theistic evolution"; see Van Till, *The Fourth Day*, p. 265. See also Numbers, *The Creationists*, pp. 177–178.

21. Henry M. Morris, "ICR and Progressive Creationism," *Acts & Facts* 24 (February 1995): 2–4; Numbers, *The Creationists*, p. 308 (Swaggart). See also Hugh Ross, *Creation and Time: A Biblical and Scientific Perspective on the Creation-Date Controversy* (Colorado Springs: NavPress, 1994).

22. Charles B. Thaxton, Walter L. Bradley, and Roger L. Olsen, *The Mystery of Life's Origin: Reassessing Current Theories* (New York: Philosophical Library, 1984), pp. vii, 186. David Berlinski, in "The Deniable Darwin," *Commentary* (June 1996): 19–29, identifies *Mathematical Challenges to the Neo-Darwinian Interpretation of Evolution*, ed. Paul S. Moorhead and Martin Kaplan (Philadelphia: Wistar Institute Press, 1967), as "the first significant criticism of evolutionary doctrine in recent decades" (p. 24).

23. Michael Denton, *Evolution: Theory in Crisis* (Bethesda, MD: Adler & Adler, 1986). See also Michael J. Denton, *Nature's Destiny: How the Laws of Biology Reveal Purpose in the Universe* (New York: Free Press, 1998).

24. Percival Davis and Dean H. Kenyon, *Of Pandas and People: The Central Question of Biological Origins*, 2nd ed. (Dallas: Haughton Publishing Co., 1993), pp. 14, 160–161; Eugenie C. Scott, "Monkey Business," *The Sciences* 36 (January/February 1996): 20–25. The "Note to Teachers" in *Of Pandas and People* was written by Mark D. Hartwig and Stephen C. Meyer.

25. Phillip E. Johnson, *Darwin on Trial* (Downers Grove, IL: InterVarsity Press, 1991); Phillip E. Johnson, *Reason in the Balance: The Case against Naturalism in Science, Law and Education* (Downers Grove, IL: InterVarsity Press, 1995), especially

pp. 15, 26. See also Phillip E. Johnson, *Defeating Darwinism by Opening Minds* (Downers Grove, IL: InterVarsity Press, 1997).

26. Michael Denton, blurb on the dust jacket of Johnson, *Darwin on Trial;* Eugenie C. Scott, "Dealing with Anti-Evolutionism," *Reports of the National Center for Science Education* 17 (July/August 1997): 24; Gould, "Impeaching a Self-Appointed Judge," pp. 118–120; Jonathan Piel to Phillip E. Johnson, June 25, 1992, mimeographed copy; Stephen C. Meyer, "A Scopes Trial for the '90s," *Wall Street Journal,* December 6, 1993; Eugenie C. Scott, "Dean Kenyon and 'Intelligent Design Theory' at San Francisco State U," *NCSE Reports* 13 (Winter 1993): 1, 5, 13. For Johnson's reaction to the treatment of Kenyon, see *Reason in the Balance,* pp. 29–30.

27. Michael J. Behe, *Darwin's Black Box: The Biochemical Challenge to Evolution* (New York: Free Press, 1996), pp. 15, 33, 193, 232–233; "The Evolution of a Skeptic: An Interview with Dr. Michael Behe, Biochemist and Author of Recent Best-Seller, *Darwin's Black Box,*" *The Real Issue* 15 (November/December 1996): 1, 6–8. In 1998 Cambridge University Press is publishing William A. Dembski's *The Design Inference: Eliminating Chance through Small Probabilities.*

28. "CT 97 Book Awards," *Christianity Today,* April 28, 1997, p. 12; David L. Wheeler, "A Biochemist Urges Darwinists to Acknowledge the Role Played by an 'Intelligent Designer,'" *Chronicle of Higher Education,* November 1, 1996, p. A13; "The Evolution of a Skeptic," pp. 7–8. See also Tom Woodward, "Meeting Darwin's Wager," *Christianity Today,* April 28, 1997, pp. 14–21.

29. William A. Dembski, "What Every Theologian Should Know about Creation, Evolution, and Design," *Transactions* 3 (May/June 1995): 1–8, published by the Center for Interdisciplinary Studies in Princeton; Eugenie C. Scott, "Old-Earth Moderates Posed to Spread Design Theory," *Reports of the National Center for Science Education* 17 (January/February 1997): 25–26; Belz, "Witness for the Prosecution," p. 18 (Goliath). See also the debate between Stephen C. Meyer ("Open the Debate on Life's Origins") and Eugenie C. Scott ("Keep Science Free from Creationism"), *Insight,* February 21, 1994, pp. 26–31. Dembski, Nelson, and Meyer are also collaborating on a forthcoming book tentatively titled *Uncommon Descent.*

30. Dembski, "What Every Theologian Should Know," p. 4; Richard Lewontin, "Billions and Billions of Demons," *New York Review of Books,* January 9, 1997, pp. 28–32, quotation on p. 31.

31. Scott Swanson, "Debunking Darwin? 'Intelligent-Design' Movement Gathers Strength," *Christianity Today,* January 6, 1997, pp. 64–65; Henry M. Morris, "Defending the Faith," *Back to Genesis* 97 (January 1997): a–c, an insert in *Acts & Facts* 26 (January 1997); and Henry M. Morris, "Neocreationism," an insert in *Acts & Facts* 27 (February 1998). See, however, Wayne Frair's positive review of *Darwin's Black Box* in the *Creation Research Society Quarterly* 34 (1997): 113.

32. J. W. Haas, Jr., "On Intelligent Design, Irreducible Complexity, and Theistic Science," *Perspectives on Science and Christian Faith* 49 (March 1997): 1. Despite his personal reservations about ID theory, Haas ran two reviews of *Darwin's Black Box,* pro and con, in the June 1997 issue of *Perspectives on Science and Christian Faith* and devoted much of the September 1997 issue to papers sympathetic to ID theory.

33. David K. Webb, Letter to the Editor, *Origins & Design* 17 (Spring 1996): 5; Berlinski, "The Deniable Darwin," pp. 19–29; "Denying Darwin: David Berlinski and Critics," *Commentary* (September 1996): 4–39, quotations on pp. 6 (Dennett) and 11 (Karl F. Wessel). See also David Berlinski, "The End of Materialist Science," *Forbes ASAP*, December 2, 1996, pp. 147–160.

34. George H. Gallup, Jr., *Religion in America 1990* (Princeton: Princeton Religion Research Center, 1990), p. 49; Garry Wills, *Under God: Religion and American Politics* (New York: Simon and Schuster, 1990), p. 124. A survey conducted in 1996 showed that 39.3 percent of American scientists believed in a personal God, down only 2.5 percent from eighty years earlier; see Edward J. Larson and Larry Witham, "Scientists Are Still Keeping the Faith," *Nature* 386 (1997): 435–436.

35. Bert James Loewenberg, "The Impact of the Doctrine of Evolution on American Thought," Ph.D. diss., Harvard University, 1934. See also Loewenberg's three published articles on Darwinism in America: "The Reaction of American Scientists to Darwinism," *American Historical Review* 38 (1933): 687–701; "The Controversy over Evolution in New England, 1859–1873," *New England Quarterly* 8 (1935): 232–257; and "Darwinism Comes to America, 1859–1900," *Mississippi Valley Historical Review* 28 (1941): 339–368.

36. A. Hunter Dupree, *Asa Gray, 1810–1888* (Cambridge: Harvard University Press, 1959); Edward Lurie, *Louis Agassiz: A Life in Science* (Chicago: University of Chicago Press, 1960); Lester D. Stephens, *Joseph LeConte: Gentle Prophet of Evolution* (Baton Rouge: Louisiana State University Press, 1982).

37. Edward J. Pfeifer, "United States," in *The Comparative Reception of Darwinism*, ed. Thomas F. Glick (Austin: University of Texas Press, 1974), pp. 168–206, which is an abstract of Pfeifer's unpublished dissertation "The Reception of Darwinism in the United States, 1859–1880," Brown University, 1957; Peter J. Bowler, "Scientific Attitudes to Darwinism in Britain and America," in *The Darwinian Heritage*, ed. David Kohn (Princeton: Princeton University Press, 1985), pp. 641–681; Mary P. Winsor, *Reading the Shape of Nature: Comparative Zoology at the Agassiz Museum* (Chicago: University of Chicago Press, 1991); Roberts, *Darwinism and the Divine in America;* James R. Moore, *The Post-Darwinian Controversies: A Study of the Protestant Struggle to Come to Terms with Darwin in Great Britain and America, 1870–1900* (Cambridge: Cambridge University Press, 1979); David N. Livingstone, *Darwin's Forgotten Defenders: The Encounter between Evangelical Theology and Evolutionary Thought* (Grand Rapids, MI: William B. Eerdmans, 1987). See also Peter Bowler, *The Eclipse of Darwinism: Anti-Darwinian Evolution Theories in the Decades around 1900* (Baltimore: Johns Hopkins University Press, 1983); and Peter Bowler, *The Non-Darwinian Revolution: Reinterpreting a Historical Myth* (Baltimore: Johns Hopkins University Press, 1988).

38. Christopher P. Toumey, *God's Own Scientists: Creationists in a Secular World* (New Brunswick, NJ: Rutgers University Press, 1994); George E. Webb, *The Evolution Controversy in America* (Lexington: University Press of Kentucky, 1994); Larson, *Trial and Error;* Numbers, *The Creationists.* See also Raymond A. Eve and Francis B. Harrold, *The Creationist Movement in Modern America* (Boston: Twayne Publishers, 1991).

39. Moore, *The Post-Darwinian Controversies*, pp. 17–122; David C. Lindberg and Ronald L. Numbers, eds., *God and Nature: Historical Essays on the Encounter between Christianity and Science* (Berkeley and Los Angeles: University of California Press, 1986); David C. Lindberg and Ronald L. Numbers, "Beyond War and Peace: A Reappraisal of the Encounter between Christianity and Science," *Church History* 55 (1986): 338–354; Ronald L. Numbers, "Science and Religion," *Osiris*, 2nd ser., 1 (1985): 59–80. The classic statement of the warfare thesis appears in Andrew Dickson White, *A History of the Warfare of Science with Theology in Christendom*, 2 vols. (New York: Appleton, 1896). For a dissenting opinion, see David A. Hollinger, "Justification by Verification: The Scientific Challenge to the Moral Authority of Christianity in Modern America," in *Religion and Twentieth-Century American Intellectual Life*, ed. Michael J. Lacey (Cambridge: Cambridge University Press, 1989), pp. 116–135.

40. For an excellent study of the changing meanings of Darwinism in Great Britain, see James Moore, "Deconstructing Darwinism: The Politics of Evolution in the 1860s," *Journal of the History of Biology* 24 (1991): 353–408.

41. Bowler, *The Non-Darwinian Revolution.*

42. See, e.g., Jon H. Roberts's exemplary account of mainstream Protestant responses to evolution, *Darwinism and the Divine in America.*

1. Darwinism and the Dogma of Separate Creations

This chapter is based on the Sarton Lecture I delivered at the annual meeting of the American Association for the Advancement of Science, Atlanta, February 17, 1995. I presented an earlier version at the annual meeting of the American Historical Association, Washington, D.C., December 28, 1992, at which time I received valuable suggestions from Deborah J. Coon and James C. Turner. More recently I have benefited greatly from the criticisms and comments of Garland E. Allen, Keith R. Benson, Peter J. Bowler, Frederick B. Churchill, Theodore J. Greenfield, Pamela M. Henson, David L. Hull, Edward J. Larson, David N. Livingstone, Jane Maienschien, Gregg A. Mitman, Lynn K. Nyhart, Philip J. Pauly, Ronald Rainger, John Stenhouse, Lester D. Stephens, Mary P. Winsor, and Suzanne E. Zeller. I owe a special debt of gratitude to my friends Jim Moore and Jon Roberts, who both know more about Darwinism than I do, and to my former graduate student Julie R. Newell, who provided invaluable research assistance. Matthew Van Atta, of the *American National Biography*, helped with biographical information, as did the numerous people acknowledged in the Appendix to this book.

1. S. S. Haldeman, "Enumeration of the Recent Freshwater Mollusca Which Are Common to North America and Europe, with Observations on Species and Their Distribution," *Journal of the Boston Society of Natural History* 4 (1844): 468–484; E. D. Cope, *The Origin of the Fittest* (New York: D. Appleton, 1887), p. 2, from "Evolution and Its Consequences," first published in *Penn Monthly Magazine* in 1872.

2. Bert James Loewenberg, "The Reaction of American Scientists to Darwinism," *American Historical Review* 38 (1933): 686–701, quotation on p. 687.

3. Edward J. Pfeifer, "United States," in The *Comparative Reception of Darwinism,* ed. Thomas F. Glick (Austin: University of Texas Press, 1974), pp. 168–206, a survey based on Pfeifer's earlier dissertation, "The Reception of Darwinism in the United States, 1859–1880," Brown University, 1957; Peter J. Bowler, "Scientific Attitudes to Darwinism in Britain and America," in *The Darwinian Heritage,* ed. David Kohn (Princeton: Princeton University Press, 1985), pp. 641–681. Of Bowler's several books on the history of evolution, the two of greatest interest to historians of American science are *The Eclipse of Darwinism: Anti-Darwinian Evolution Theories in the Decades around 1900* (Baltimore: Johns Hopkins University Press, 1983), and *The Non-Darwinian Revolution: Reinterpreting a Historical Myth* (Baltimore: Johns Hopkins University Press, 1988). Other valuable studies of scientific responses to Darwin include A. Hunter Dupree, *Asa Gray, 1810–1888* (Cambridge: Harvard University Press, 1959); Edward Lurie, *Louis Agassiz: A Life in Science* (Chicago: University of Chicago Press, 1960); Lester D. Stephens, *Joseph LeConte: Gentle Prophet of Evolution* (Baton Rouge: Louisiana State University Press, 1982); and Mary P. Winsor, *Reading the Shape of Nature: Comparative Zoology at the Agassiz Museum* (Chicago: University of Chicago Press, 1991), which examines the responses of Louis Agassiz's students to evolution. For a dated, but still useful, introduction to the literature, see Michele L. Aldrich, "United States: Bibliographical Essay," in Glick, ed., *The Comparative Reception of Darwiniam,* pp. 207–226.

4. On the creation of the National Academy of Sciences, see A. Hunter Dupree, *Science in the Federal Government: A History of Policies and Activities to 1940* (Cambridge: Harvard University Press, 1957), pp. 135–148.

5. James Moore, "Deconstructing Darwinism: The Politics of Evolution in the 1860s," *Journal of the History of Biology* 24 (1991): 353–408; David L. Hull, "Darwinism As a Historical Entity: A Historiographic Proposal," in *The Darwinian Heritage,* ed. Kohn, pp. 773–812; Ernst Mayr, *One Long Argument: Charles Darwin and the Genesis of Modern Evolutionary Thought* (Cambridge: Harvard University Press, 1991), pp. 90–107.

6. Charles Darwin, *The Descent of Man, and Selection in Relation to Sex,* 2 vols. (London: John Murray, 1871), 1:152–153; Charles Darwin to Asa Gray, May 11, 1863, quoted in Francis Darwin, ed., *The Life and Letters of Charles Darwin,* 2 vols. (New York: D. Appleton, 1896), 2:163–164.

7. On American usage, see Charles Hodge, *What Is Darwinism?* (New York: Scribner, Armstrong, 1874), p. 104.

8. Asa Gray to J. D. Hooker, January 5, 1860 (fair play), quoted in *The Life and Letters of Charles Darwin,* 2:63; Asa Gray to Charles Darwin, January 23, 1860 (organs), quoted ibid., 2:66; Asa Gray, *Darwiniana: Essays and Reviews Pertaining to Darwinism,* ed. A. Hunter Dupree (Cambridge: Harvard University Press, 1963), pp. 5 (a Darwinian), 77 (special origination); Asa Gray to J. D. Dana, June 22, 1872, quoted in Daniel C. Gilman, *The Life of James Dwight Dana* (New York: Harper and Brothers, 1899), pp. 301–302; Hodge, *What Is Darwinism?* pp. 174–175. By 1880 Gray no longer regarded "man's origin as exceptional in the sense of directly supernatural"; Asa Gray, *Natural Science and Religion: Two Lectures Delivered to the Theological School of Yale College* (New York: Charles Scribner's Sons, 1880), p. 99. On Darwin's rejection of Gray's theistic evolution, see Charles Darwin, *The Variation of*

Animals and Plants under Domestication, 2 vols. (New York: D. Appleton, 1897), 2:428, first published in 1868. On Gray as the prototypical American Darwinian, see, e.g., James R. Moore, *The Post-Darwinian Controversies: A Study of the Protestant Struggle to Come to Terms with Darwin in Great Britain and America, 1870–1900* (Cambridge: Cambridge University Press, 1979), pp. 269–272; and George Cotkin, *Reluctant Modernism: American Thought and Culture, 1880–1900* (New York: Twayne Publishers, 1992), p. 8. On Gray's life and work, see especially Dupree, *Asa Gray.*

9. Andrew Denny Rogers III, *American Botany, 1873–1892: Decades of Transition* (Princeton: Princeton University Press, 1944), pp. 5, 19, 190–191; *The Life and Letters of Charles Darwin,* 2:216. For biographical accounts of Lesquereux, see Edward Orton, "Leo Lesquereux," *American Geologist* 5 (1890): 284–296; and J. P. Lesley, "Memoir of Leo Lesquereux, 1806–1889," *Biographical Memoirs* (National Academy of Sciences) 3 (1895): 187–212.

10. Charles Darwin, *On the Origin of Species,* with an introduction by Ernst Mayr (Cambridge: Harvard University Press, 1966), p. 437; Tore Frängsmyr, "Linnaeus as a Geologist," in *Linnaeus: The Man and His Work,* ed. Tore Frängsmyr (Berkeley and Los Angeles: University of California Press, 1983), pp. 110–155, especially p. 122; Charles Lyell, *Principles of Geology,* 3 vols. (London: John Murray, 1830–1833), 2:123–126; Louis Agassiz, *Essay on Classification,* ed. Edward Lurie (Cambridge: Harvard University Press, 1962), pp. 173–175; Gray, *Natural Science and Religion,* p. 35. Regarding Agassiz's views, see also Ernst Mayr, "Agassiz, Darwin, and Evolution," *Harvard Library Bulletin* 13 (1959): 165–194; and Mary P. Winsor, "Louis Agassiz and the Species Question," *Studies in History of Biology* 3 (1979): 89–117. On notions of special creation in the nineteenth century, see Chapter 2, as well as Neal C. Gillespie, *Charles Darwin and the Problem of Creation* (Chicago: University of Chicago Press, 1979), chap. 2; and Ernst Haeckel, *The History of Creation,* trans. E. Ray Lankester, 2 vols. (New York: D. Appleton, 1880), 1:24–71.

11. Lurie, *Louis Agassiz,* p. 262 (church); D[avid] S[tarr] J[ordan] and J[essie] K[night] J[ordan], "Louis Agassiz," *Dictionary of American Biography,* 1:114–122, quotation on p. 122 (infidel); Agassiz, *Essay on Classification,* p. 95 (deluge); Dorothy G. Wyman, *Edward Sylvester Morse: A Biography* (Cambridge: Harvard University Press, 1942), p. 120 (Adam and Eve); Ronald L. Numbers, *The Creationists* (New York: Alfred A. Knopf, 1992), p. 8 (tenet teaching). On Agassiz's career, see also Jules Marcou, *Life, Letters, and Works of Louis Agassiz,* 2 vols. (New York: Macmillan, 1896).

12. Joseph LeConte, "Memoir of James Dwight Dana," *Bulletin of the Geological Society of America* 7 (1896): 461–479, quotation (prepared the way) on p. 468; Mayr, "Agassiz, Darwin, and Evolution," p. 179 (order of succession); Charles Darwin to J. D. Hooker, March 26, 1854, quoted in *The Life and Letters of Charles Darwin,* 1:403; A. S. Packard, Jr., "The Philosophical Views of Agassiz," *American Naturalist* 32 (1898): 161–163 (great fellow), quoted in Ralph W. Dexter, "The Impact of Evolutionary Theories on the Salem Group of Agassiz Zoologists (Morse, Hyatt, Packard, Putnam)," *Essex Institute Historical Collections* 115 (1979): 144–171, quotation on p. 146. On Agassiz's positive contribution to the Darwinian debates in America, see also Edward S. Morse, "Address," *Proceedings of the American Association for the Advancement of Science* 25 (1876): 136–176, especially pp. 138–140. The

best discussion of Agassiz's students appears in Mary P. Winsor, *Reading the Shape of Nature: Comparative Zoology at the Agassiz Museum* (Chicago: University of Chicago Press, 1991).

13. [William Hayes Ward], "Do Our Colleges Teach Evolution?" *Independent* 31 (December 18, 1879): 14–15; Arnold Guyot, *Creation; or, The Biblical Cosmogony in the Light of Modern Science* (New York: Charles Scribner's Sons, 1884), pp. 116–128; James D. Dana, "Memoir of Arnold Guyot," *Biographical Memoirs* (National Academy of Sciences) 2 (1886): 309–347, quotation on p. 334. For Guyot's views on the nebular hypothesis, see Ronald L. Numbers, *Creation by Natural Law: Laplace's Nebular Hypothesis in American Thought* (Seattle: University of Washington Press, 1977), pp. 91–100. On Guyot's skeptical attitude toward Darwinism, see also his letter to O. C. Marsh, October 16, 1877, quoted in Charles Schuchert and Clara Mae LeVene, *O. C. Marsh: Pioneer in Paleontology* (New Haven: Yale University Press, 1940), p. 243. Despite the *Independent*'s claim, it seems likely that at least two other American naturalists of repute, George Engelmann and J. P. Lesley, continued to resist evolution.

14. Gray, *Darwiniana*, p. 77; J. D. Dana to John G. Hall, March 7, 1889, quoted in Michael L. Prendergast, "James Dwight Dana: The Life and Thought of an American Scientist," Ph.D. diss., University of California, Los Angeles, 1978, p. 516. See also James D. Dana, "Views of Evolution," *Independent* 32 (January 8, 1880): 2–3; and James D. Dana, *Manual of Geology*, 4th ed. (New York: American Book Co., 1895), p. 1036. On Morgan, see Carl Resek, *Lewis Henry Morgan: American Scholar* (Chicago: University of Chicago Press, 1960), pp. 96, 99–100, 129; Adam Kuper, "The Development of Lewis Henry Morgan's Evolutionism," *Journal of the History of the Behavioral Sciences* 21 (1985): 3–22, especially p. 4; and Thomas R. Trautmann, *Lewis Henry Morgan and the Invention of Kinship* (Berkeley and Los Angeles: University of California Press, 1987), pp. 32, 174–175. On Putnam, see Dexter, "The Impact of Evolutionary Theories on the Salem Group," pp. 169–171.

15. John M. Clarke, *James Hall of Albany: Geologist and Paleontologist, 1811–1898* (Albany, NY: n.p., 1921), p. 508; Mary Lesley Ames, ed., *Life and Letters of Peter and Susan Lesley*, 2 vols. (New York: G. P. Putnam's Sons, 1909), 2:473; W. M. Davis, "Biographical Memoir of Peter Lesley, 1819–1903," *Biographical Memoirs* (National Academy of Sciences) 8 (1919): 153–240, quotation on p. 225.

16. A study of sixty-seven British scientists concluded that "older scientists were as quick to change their minds as younger scientists"; David L. Hull, Peter D. Tessner, and Arthur M. Diamond, "Planck's Principle: Do Younger Scientists Accept New Scientific Ideas with Greater Alacrity than Older Scientists?" *Science* 202 (1978): 717–723, quotation on p. 722. For a contrasting view, see Michael Ruse, *The Darwinian Revolution: Science Red in Tooth and Claw* (Chicago: University of Chicago Press, 1979), p. 229. See also S. G. Levin, P. E. Stephan, and M. B. Walker, "Planck's Principle Revisited: A Note," *Social Studies of Science* 25 (1995): 275–283.

17. Charles Darwin to Asa Gray, July 20, 1857, quoted in Dupree, *Asa Gray*, pp. 244–245; Charles Darwin to Joseph Leidy, March 4, 1860, quoted in Henry Fairfield Osborn, "Darwin and Paleontology," in *Fifty Years of Darwinism*, by the American Association for the Advancement of Science (New York: Henry Holt, 1909), pp. 209–250, quotation (regarding earlier promise of support) on pp. 209–

210; Ronald Rainger, *An Agenda for Antiquity: Henry Fairfield Osborn and Vertebrate Paleontology at the American Museum of Natural History, 1890–1935* (Tuscaloosa: University of Alabama Press, 1991), p. 11 (meteor); Henry Fairfield Osborn, "Biographical Memoir of Joseph Leidy, 1823–1891," *Biographical Memoirs* (National Academy of Sciences) 7 (1913): 335–396, especially p. 368 (regarding election); Charles Darwin to Charles Lyell, May 8, 1860, quoted in *The Life and Letters of Charles Darwin,* 2:100 (appreciation).

18. Dupree, *Asa Gray,* pp. 285–288; *Proceedings of the Boston Society of Natural History* 7 (1859–1861): 231–233, 248; William Barton Rogers to Henry Darwin Rogers, March 20, 1860, quoted in Emma Rogers, *Life and Letters of William Barton Rogers,* 2 vols. (Boston: Houghton Mifflin, 1896), 2:29–30. W. M. Smallwood reported that he searched the Boston newspapers in vain for any mention of the Agassiz-Rogers debate; see W. M. Smallwood, "The Agassiz-Rogers Debate on Evolution," *Quarterly Review of Biology* 16 (1941): 1–12, especially p. 11. However, on March 5, 1860, the *Boston Courier* did carry a positive review of the *Origin,* a clipping of which can be found in the William Barton Rogers Collection (MC 1), MIT Archives. I am indebted to Loretta H. Mannix for providing me with a copy of this review.

19. John F. Fulton and Elizabeth H. Thomson, *Benjamin Silliman, 1779–1864: Pathfinder in American Science* (New York: Henry Schuman, 1947), p. 258; Andrew Denny Rogers III, *John Torrey: A Story of North American Botany* (Princeton: Princeton University Press, 1942), p. 278; "The Advance of Science," *New York Times,* August 6, 1860; "Progress and Prospects of Science in America," ibid., August 25, 1860; J. S. Newberry, "Address," *Proceedings of the American Association for the Advancement of Science* 16 (1867): 1–15; quotations on pp. 10–11. Regarding Torrey, see also Christine Chapman Robbins, "John Torrey (1796–1873): His Life and Times," *Bulletin of the Torrey Botanical Club* 95 (1968): 515–645, especially p. 632. The Illinois entomologist Benjamin D. Walsh was among the earliest advocates of evolution in America; see Winsor, *Reading the Shape of Nature,* pp. 94, 99–100; and Pfeifer, "United States," p. 184.

20. F. W. Putnam to Adelaide Edmands, July 13, 1860, quoted in Dexter, "The Impact of Evolutionary Theories on the Salem Group," p. 169 (Agassiz); Alpheus Hyatt to J. W. Dawson, March 31, 1870, Box 3, Dawson Collection, McLennan Library, McGill University; W. G. Farlow, "The Conception of Species as Affected by Recent Investigations on Fungi," *American Naturalist* 32 (1898): 675–696, quotation on p. 678. On early philosophical discussions of Darwinism in Cambridge, see Philip P. Wiener, *Evolution and the Founders of Pragmatism* (Cambridge: Harvard University Press, 1949). See also W. G. Farlow, "The Change from the Old to the New Botany in the United States," *Science,* n.s., 37 (1913): 79–86. Regarding the teaching of Darwinism in American colleges, see David Edward Ellen, "Darwin in the Nineteenth Century Harvard Classroom: From the Old to the New Curriculum," B.A. thesis, Harvard University, 1987; and Clifford Harold Peterson, "The Incorporation of the Basic Evolutionary Concepts of Charles Darwin in Selected American College Biology Programs in the Nineteenth Century," Ed.D. diss., Columbia University, 1970, which concludes that Darwinism "did not appear in college biology courses and textbooks until the 1870's," p. 189.

21. J. D. Whitney, *Lecture on Geology: Delivered before the Legislature of California, at San Francisco, Thursday Evening, Feb. 27, 1862* (San Francisco: Benj. P. Avery, 1862), pp. 24–25; Alpheus Hyatt, "On the Parallelism between the Different States of Life in the Individual and Those in the Entire Group of the Molluscous Order Tetrabrachiata," *Memoirs of the Boston Society of Natural History* 1 (1866): 193–209; Hyatt's untitled comments in the *Proceedings of the Boston Society of Natural History* 14 (1871): 148; Winsor, *Reading the Shape of Nature,* pp. 41–42; Edward D. Cope, "On the Origin of Genera," *Proceedings of the Academy of Natural Sciences of Philadelphia* 20 (1868): 242–300; E. S. Morse's untitled remarks in the *Bulletin of the Essex Institute* 7 (1874): 138; Louis Agassiz, "Evolution and the Permanence of Types," *Atlantic Monthly* 33 (1874): 95. Regarding Whitney's experience, see Edwin Tenney Brewster, *Life and Letters of Josiah Dwight Whitney* (Boston: Houghton Mifflin, 1909), pp. 199–300. F. V. Hayden and F. B. Meek, for example, accepted Darwinism at some point, but according to Hayden's biographer, "they remained silent on this subject in their printed works and even in their correspondence"; Mike Foster, *Strange Genius: The Life of Ferdinand Vandeveer Hayden* (Niwot, CO: Roberts Rinehart, 1994), p. 110.

22. Darwin, *On the Origin of Species,* p. 466. Similar statements appear in the first edition on pp. 29, 82, 85, 132–133, 136, 143, 167–168, and 209. On Darwin's changing views, see Peter J. Vorzimmer, *Charles Darwin: The Years of Controversy: The "Origin of Species" and Its Critics, 1859–1882* (Philadelphia: Temple University Press, 1970).

23. Gray, *Darwiniana,* pp. 121–122; W. B. Rogers to H. D. Rogers, January 2, 1860, and December 1, 1861, quoted in Rogers, *Life and Letters of William Barton Rogers,* 2:18–19, 104; O. C. Marsh, "History and Methods of Paleontological Discovery," *American Journal of Science* 118 (1879): 323–359, quotation on p. 351 (magic word); Bowler, *The Non-Darwinian Revolution,* p. 80 (regarding Marsh).

24. Alpheus S. Packard, *Lamarck, the Founder of Evolution: His Life and Work* (New York: Longmans, Green, 1901), pp. 395–396; A. S. Packard, Jr., "A Century's Progress in American Zoology," *American Naturalist* 10 (1876): 591–598, quotation on p. 597; Theodore J. Greenfield, "Variation, Heredity, and Scientific Explanation in the Evolutionary Theory of Four American Neo-Lamarckians, 1867–1897," Ph.D. diss., University of Wisconsin–Madison, 1986, quotation on p. 24. For early assessments of neo-Lamarckism, see George W. Stocking, Jr., "Lamarckianism in American Social Science, 1890–1915," *Journal of the History of Ideas* 23 (1962): 239–256; and Edward J. Pfeifer, "The Genesis of American Neo-Lamarckism," *Isis* 56 (1965): 221–231. For an example of the difficulty even contemporaries had in defining neo-Lamarckism, see J. T. Cunningham, "Lyell and Lamarckism: A Reply to Professor W. K. Brooks," *Natural Science* 8 (1896): 326–331; and W. K. Brooks, "Lyell and Lamarckism: A Rejoinder," ibid. 9 (1896): 115–119.

25. Packard, *Lamarck,* pp. 400–401; E. D. Cope, "A Review of the Modern Doctrine of Evolution," *American Naturalist* 14 (1880): 166–179, 260–271, quotation on p. 175; E. S. Morse, "Address," *Proceedings of the American Association for the Advancement of Science* 36 (1887): 1–43, quotation on pp. 16–17. On the neo-Lamarckians' search for natural, external causes, see Greenfield, "Variation, Heredity, and Scientific Explanation," pp. 14, 26, 54–58.

26. Vernon L. Kellogg, *Darwinism To-Day* (New York: Henry Holt, 1907), pp. 285–286 (bathmism, etc.); Henry Fairfield Osborn, *Cope: Master Naturalist: The Life and Letters of Edward Drinker Cope* (Princeton: Princeton University Press, 1931), p. 533 (Catagenesis, etc.); William Keith Brooks, "Biographical Memoir of Alpheus Hyatt, 1838–1902," *Biographical Memoirs* (National Academy of Sciences) 6 (1909): 311–325, quotation on p. 321; Charles Darwin to E. S. Morse, April 23, 1877, quoted in *The Life and Letters of Charles Darwin,* 2:409.

27. The most extensive contemporary list of neo-Lamarckians I have seen appears in E. D. Cope, *The Primary Factors of Organic Evolution* (Chicago: Open Court, 1896), pp. 8–9, 521–526, which names the academicians Beecher, Cope, Dall, Hyatt, Osborn, and Packard, as well as D. G. Elliot, Robert T. Jackson, Henry B. Orr, C. V. Riley, John A. Ryder, W. B. Scott, Benjamin Sharp, Lester Frank Ward, and Jacob Wortman. Kellogg, *Darwinism To-Day,* pp. 272–305, adds Luther Burbank and Charles L. Redfield to the list; while Theodore Gill identifies himself as a somewhat reluctant neo-Lamarckian in "Edward Drinker Cope," p. 240. In 1876 Packard cited the work of the academicians Joel A. Allen, Spencer F. Baird, O. C. Marsh, and Edward S. Morse (along with that of Cope, Hyatt, Gill, Packard, and the nonacademician Robert Ridgway) in documenting the existence of "an original and distinctively American school of evolutionists," but at the same time he included Morse in the opposing camp of natural selectionists, where Marsh, too, might have better fit; Packard, "A Century's Progress in American Zoology," p. 597. The British naturalist George J. Romanes identified William Keith Brooks as being among "the most prominent" of the American neo-Lamarckians (with Cope, Hyatt, Packard, Ryder, Dall, and Osborn), a characterization that prompted Brooks to issue a correction. See Romanes, *Darwin, and after Darwin,* 2:14; and W. K. Brooks, "Lamarck and Lyell: A Short Way with Lamarckians," *Natural Science* 8 (1896): 89–93. Early in the twentieth century William North Rice recalled that his mentor, James Dwight Dana, had held "more nearly a neo-Lamarckian than a strictly Darwinian view of the method of evolution"; see Rice, "James Dwight Dana, Geologist (1813–1895)," in *Leading American Men of Science,* ed. David Starr Jordan (New York: Henry Holt, 1910), pp. 233–268, quotation on p. 250. On the alleged popularity of neo-Lamarckism, see, e.g., Kellogg, *Darwinism To-Day,* p. 157; Pfeifer, "United States," p. 199; and George E. Webb, *The Evolution Controversy in America* (Lexington: University Press of Kentucky, 1994), p. 28. Regarding changing attitudes toward the inheritance of acquired characters, see Greenfield, "Variation, Heredity, and Scientific Explanation," p. 188 (Cope, Hyatt, Packard); Gill, "Edward Drinker Cope," pp. 240–241 (Gill); and Rainger, *An Agenda for Antiquity,* p. 40 (Osborn).

28. Kellogg, *Darwinism To-Day,* p. 188 (all-sufficient); Lester F. Ward, "Neo-Darwinism and Neo-Lamarckism," *Proceedings of the Biological Society of Washington* 6 (1890–91): 11–71, especially p. 51 (Romanes). On Brown-Séquard's research, see J. M. D. Olmsted, *Charles-Edouard Brown-Séquard: A Nineteenth Century Neurologist and Endocrinologist* (Baltimore: Johns Hopkins University Press, 1946); and Michael J. Aminoff, *Brown-Séquard: A Visionary of Science* (New York: Raven Press, 1993). On Darwin's reaction, see A. Hunter Dupree, ed., "Some Letters from Charles Darwin to Jeffries Wyman," *Isis* 42 (1951): 104–110, especially p. 106. On Weis-

mann, see Frederick B. Churchill, "August Weismann and a Break from Tradition," *Journal of the History of Biology* 1 (1968): 91–112; Frederick B. Churchill, "The Weismann-Spencer Controversy over the Inheritance of Acquired Characters," in *Human Implications of Scientific Advance,* ed. E. G. Forbes (Edinburgh: Edinburgh University Press, 1978), pp. 451–468, which briefly discusses efforts to replicate Brown-Séquard's experiments; Greenfield, "Variation, Heredity, and Scientific Explanation," passim; and Mayr, *One Long Argument,* pp. 108–131. See also L. I. Blacher, *The Problem of the Inheritance of Acquired Characters,* trans. F. B. Churchill (New Delhi: Amerind Publishing Co., 1982). On circumcision and the noninheritance of acquired characters, see Charles Darwin to Jeffries Wyman, October 3, 1860, quoted in Dupree, ed., "Some Letters from Charles Darwin to Jeffries Wyman," pp. 106–107; and Theodore Gill, "Edward Drinker Cope, Naturalist: A Chapter in the History of Science," *Science,* n.s., 6 (1897): 225–243, especially p. 241.

29. Ward, "Neo-Darwinism and Neo-Lamarckism," p. 33; Bowler, *The Eclipse of Darwinism,* p. 42.

30. David Starr Jordan and Vernon Lyman Kellogg, *Evolution and Animal Life* (New York: D. Appleton, 1907), p. 197, which Keith Benson kindly brought to my attention; Keith Rodney Benson, "William Keith Brooks (1848–1908): A Case Study in Morphology and the Development of American Biology," Ph.D. diss., Oregon State University, 1979, p. 239; William K. Brooks, *The Law of Heredity: A Study of the Cause of Variation and the Origin of Living Organisms* (Baltimore: J. Murphy, 1883), p. 164, quoted ibid., p. 245; Brooks, "Lamarck and Lyell," p. 89. Dupree, *Asa Gray,* the standard biography of the Harvard botanist, mentions neither Weismann nor neo-Darwinism.

31. Edmund B. Wilson, *The Cell in Development and Inheritance* (New York: Macmillan, 1896), p. 11 (Weismann); Garland E. Allen, "Edmund Beecher Wilson," *Dictionary of Scientific Biography,* 14:423–436, quotations on p. 434; Edmund B. Wilson, Introduction to Charles Darwin, *The Origin of Species by Means of Natural Selection* (New York: Macmillan, 1927), p. xiii (organs); H. J. Muller, "Edmund B. Wilson—An Appreciation," *American Naturalist* 77 (1943): 142–172, quotation on pp. 153–154 (higgledy-piggledy). See also T. H. Morgan, "Biographical Memoir of Edmund Beecher Wilson, 1856–1939," *Biographical Memoirs* (National Academy of Sciences) 21 (1941): 315–342; Alice Baxter Levine, "Edmund Beecher Wilson and the Problem of Development: From the Germ Layer Theory to the Chromosome Theory of Inheritance," Ph.D. diss., Yale University, 1974; and Jane Maienschein, *Transforming Traditions in American Biology, 1880–1915* (Baltimore: Johns Hopkins University Press, 1991). John Herschel was apparently the first to describe evolution by natural selection as "the law of higgledy-piggledy"; see Charles Darwin to Charles Lyell, December 12, 1859, in *The Life and Letters of Charles Darwin,* 2:37.

32. According to Frederick B. Churchill (personal communication, October 23, 1992), Weismann's American students included Howard Ayers, Harris Hawthorne Wilder, and Henry Lane Bruner. Carl H. Eigenmann, an Indiana zoologist elected to the Academy in 1923, spent a semester in Weismann's laboratory. On the popularity of neo-Darwinism in England, see Kellogg, *Darwinism To-Day,* p. 157;

and Bowler, The *Eclipse of Darwinism,* p. 42. Recently Pamela M. Henson has portrayed the Cornell entomologist J. H. Comstock, a nonmember of the Academy, as "an ardent Darwinian selectionist." Comstock did consider himself a Darwinist, but in his major statement on evolution he referred positively to Hyatt's views and adopted an agnostic attitude on the *origin* of variations. Like so many of his contemporaries, he invoked natural selection primarily to explain the elimination of the unfit. Pamela M. Henson, "Evolution and Taxonomy: J. H. Comstock's Research School in Evolutionary Entomology at Cornell University, 1874–1930," Ph.D. diss., University of Maryland, 1990, p. 159; John Henry Comstock, "Evolution and Taxonomy," in *The Wilder Quarter-Century Book* (Ithaca, NY: Comstock Publishing Co., 1893), pp. 37–113, especially pp. 47, 51–53. See also Pamela M. Henson, "The Comstock Research School in Evolutionary Entomology," *Osiris,* n.s., 8 (1993): 159–177. According to W. Conner Sorensen, the American entomologist William Henry Edwards not only collaborated with Weismann from 1875 to 1878 but contributed to his rejection of Lamarckism; however, Sorensen does not say whether or not Edwards adopted Weismann's views on natural selection. See Sorensen, *Brethren of the Net: American Entomology, 1840–1880* (Tuscaloosa: University of Alabama Press, 1995), chap. 10.

33. William H. Brewer, "On the Hereditary Transmission of Acquired Characters," *Agricultural Science* 6 (1892): 103–107, 153–156, 249–254, 345–348, especially p. 105; Charles Sedgwick Minot, "The Physical Basis of Heredity," *Science* 8 (1886): 125–130, especially p. 129; Joseph LeConte, *Evolution: Its Nature, Its Evidences, and Its Relation to Religious Thought,* 2nd ed. (New York: D. Appleton, 1891), p. 97, which Keith Benson brought to my attention; Henry Fairfield Osborn, quoted in Rainger, *An Agenda for Antiquity,* p. 125. See also Henry Fairfield Osborn, "The Hereditary Mechanism and the Search for the Unknown Factors of Evolution," *American Naturalist* 23 (1895): 418–439.

34. Edward S. Morse, "Address," *Proceedings of the American Association for the Advancement of Science* 25 (1876): 136–176, especially p. 176; Morse, "Address," ibid. 36 (1887): 1–43, quotation on p. 2; J. W. Powell, "From Barbarism to Civilization," *American Anthropologist* 1 (1888): 97–123, quotation on p. 122. See also Powell, "Human Evolution," *Transactions of the Anthropological Society of Washington* 2 (1883): 176–208. Significantly, Morse invoked Darwin, not Spencer, in defense of his views.

35. The sixteen presumed evolutionists for whom I have found insufficient evidence to classify their views on the mechanism of evolution are Spencer F. Baird, C.-E. Brown-Séquard, William G. Farlow, Grove Karl Gilbert, George Brown Goode, S. S. Haldeman, F. V. Hayden, Joseph Leidy, Leo Lesquereux, F. B. Meek, Lewis Henry Morgan, Frederic W. Putnam, Sydney I. Smith, William Stimpson, William H. Welch, and Amos Henry Worthen. The eight quasi-Lamarckians are Joel A. Allen, William Henry Brewer, Elliott Coues, Clarence E. Dutton, George L. Goodale, Charles S. Minot, John Wesley Powell, and Addison E. Verrill. I have a hard time pinning Edward S. Morse down, but he, too, might fit into this category, as might James Dwight Dana, a theistic evolutionist.

36. Clarence King, "Catastrophism and Evolution," *American Naturalist* 11 (1877): 449–470, quotations on pp. 464, 470; Joseph LeConte, "On Critical Periods

in the History of the Earth and Their Relation to Evolution," *American Journal of Science* 114 (1877): 99–114, quotation on p. 101; Joseph LeConte, The *Autobiography of Joseph LeConte,* ed. William Dallam Armes (New York: D. Appleton, 1903), p. 266 (most important); G. Frederick Wright, "Present Aspects of the Questions concerning the Origin and Antiquity of the Human Race," *Protestant Episcopal Review* 11 (1898): 319. On the attraction of the neo-catastrophists to the mutation theory, see Hans Jenny, *E. W. Hilgard and the Birth of Modern Soil Science* (Pisa: Collana Della Revista, 1961), pp. 113–114; Charles A. White, "The Saltatory Origin of Species," *Bulletin of the Torrey Botanical Club* 29 (1902): 511–522; and Charles A. White, "The Mutation Theory of Professor De Vries," *Annual Report of the Smithsonian Institution for 1901,* 2 vols., 1:631–640. On the mutation theory at the turn of the century, see Peter J. Bowler, The *Eclipse of Darwinism: Anti-Darwinian Evolution Theories in the Decades around 1900* (Baltimore: Johns Hopkins University Press, 1983), pp. 182–226. The six catastrophists are Eugene W. Hilgard, Clarence King, Joseph LeConte, William Barton Rogers, Charles D. Walcott, and Charles A. White. The neo-Lamarckian William H. Dall could also be added to this group.

37. Charles Hartshorne and Paul Weiss, *Collected Papers of Charles Sanders Peirce,* vol. 6: *Scientific Metaphysics* (Cambridge: Harvard University Press, 1935), pp. 16–17.

38. Charles Otis Whitman, "The Problem of the Origin of Species," in *Congress of Arts and Science: Universal Exposition, St. Louis, 1904,* ed. Howard J. Rogers, 8 vols. (Boston: Houghton Mifflin, 1905–1907), 5:41–58, quotations on pp. 44–45, 57; Edward Beecher Wilson, *Biology* (New York: Columbia University Press, 1908), p. 22. Regarding Whitman, see Edward S. Morse, "Biographical Memoir of Charles Otis Whitman, 1842–1910," *Biographical Memoirs* (National Academy of Sciences) 7 (1913): 269–288; Ernst Mayr, "Charles Otis Whitman," *Dictionary of Scientific Biography* 14:313–315; and Philip J. Pauly, "From Adventism to Biology: The Development of Charles Otis Whitman," *Perspectives in Biology and Medicine* 37 (1994): 395–408. The anthropologist Franz Boas also stressed the importance of "inherent forces"; see his autobiographical essay in Clifton Fadiman, ed., *I Believe* (New York: Simon and Schuster, 1939), pp. 19–29, quotation on p. 29. On the history of orthogenetic theories, see Bowler, *The Eclipse of Darwinism,* pp. 141–181.

39. Samuel Hubbard Scudder, *The Butterflies of the Eastern United States and Canada, with Special Reference to New England,* 3 vols. (Cambridge: By the Author, 1889), 2:953; Editor of the *Independent,* quoted in Michael McGiffert, "Christian Darwinism: The Partnership of Asa Gray and George Frederick Wright, 1874–1881," Ph.D. diss., Yale University, 1958, p. 229; Bowler, *The Eclipse of Darwinism,* p. 44; S. E. Coues, *Studies of the Earth* (1860), quoted in "The Logical Relations of Religion and Natural Science," *Princeton Review* 32 (1860): 578. On Gray's failure as a Darwinian theist, see also Dupree, *Asa Gray,* p. 381; W. G. Farlow, "Memoir of Asa Gray, 1810–1888," *Biographical Memoirs* (National Academy of Sciences) 3 (1895): 161–175, especially p. 172; and George Macloskie, quoted in William Milligan Sloane, ed., *The Life of James McCosh: A Record Chiefly Autobiographical* (New York: Charles Scribner's Sons, 1897), p. 123. Gray's leading disciple was the geologist-cleric George Frederick Wright; see Ronald L. Numbers, "George Frederick Wright: From Christian Darwinist to Fundamentalist," *Isis* 79 (1988): 624–645.

As late as 1926 Edwin E. Slosson, the knowledgeable director of Science Service, thought that "it would be hard, if not impossible, to find any man of standing who would argue for evolution based on materialistic and atheistic principles"; Edwin E. Slosson to "Joint Editors," October 18, 1926, Kirtley F. Mather Papers, Harvard University Archives. The eight theists are Henry James Clark, James Dwight Dana, Asa Gray, J. L. LeConte, Samuel H. Scudder, J. D. Whitney, J. J. Woodward, and Jeffries Wyman. Clark adopted the views of the British naturalist Richard Owen; see A. S. Packard, Jr., "Memoir of Henry James Clark, 1826–1873," *Biographical Memoirs* (National Academy of Sciences) 1 (1877): 319–328, especially p. 323.

40. Bowler, *The Eclipse of Darwinism,* p. 27.

41. James R. Moore, "Of Love and Death: Why Darwin 'Gave up Christianity,' " in *History, Humanity, and Evolution: Essays for John C. Greene,* ed. James R. Moore (Cambridge: Cambridge University Press, 1989), pp. 195–229.

42. Jon H. Roerts, *Darwinism and the Divine in America: Protestant Intellectuals and Organic Evolution, 1859–1900* (Madison: University of Wisconsin Press, 1988), p. 136; Joseph LeConte, "Lectures on Coal," Smithsonian Institution, *Annual Report* (1858), pp. 119–168, quotation on p. 168, for which I am indebted to Lester D. Stephens; Joseph LeConte, "Evolution in Relation to Materialism," *Princeton Review,* n.s., 7 (1881): 149–174, quotations on pp. 166, 174; A. Hunter Dupree, "Christianity and the Scientific Community in the Age of Darwin," in *God and Nature: Historical Essays on the Encounter between Christianity and Science,* ed. David C. Lindberg and Ronald L. Numbers (Berkeley and Los Angeles: University of California Press, 1986), pp. 351–368, quotation on p. 352. The best introduction to theological responses to Darwinism in America is Roberts, *Darwinism and the Divine in America,* but see also Moore, *The Post-Darwinian Controversies;* Numbers, *The Creationists;* and David N. Livingstone, *Darwin's Forgotten Defenders: The Encounter between Evangelical Theology and Evolutionary Thought* (Grand Rapids, MI: William B. Eerdmans, 1987).

43. E. S. Morse to John Gould, March 11, 1860, and October 25, 1873, quoted in Dorothy G. Wyman, *Edward Sylvester Morse: A Biography* (Cambridge: Harvard University Press, 1942), pp. 225–226, and ibid., pp. 22, 52; [William Hayes Ward], "Christian Evolution," *Independent* 32 (January 8, 1880): 16–17.

44. Moore, *The Post-Darwinian Controversies,* p. 109; William F. Sanford, Jr., "Dana and Darwinism," *Journal of the History of Ideas* 26 (1965): 531–546, quotations on pp. 531, 543; A. Hunter Dupree, "Jeffries Wyman's Views on Evolution," *Isis* 44 (1953): 243–246, quotation on p. 245 (distress); Toby A. Appel, "Jeffries Wyman, Philosophical Anatomy, and the Scientific Reception of Darwin in America," *Journal of the History of Biology* 21 (1988): 69–94, quotation on p. 71 (little difficulty); Joseph LeConte, *Religion and Science: A Series of Sunday Lectures on the Relation of Natural and Revealed Religion; or, The Truths Revealed in Nature and Scripture* (New York: D. Appleton, 1873), p. 276. Dana's friend Arnold Guyot did on one occasion express concern that the public debate over Dana's views on evolution was causing him emotional distress; see Arnold Guyot to Mrs. J. D. Dana, January 17, 1880, and Arnold Guyot to J. D. Dana, February 16, 1880, James Dwight Dana Correspondence, Yale University Library.

45. Moore, *The Post-Darwinian Controversies*, p. 303.

46. Jon H. Roberts has recently made the same point for Protestant thinkers generally; see his "Darwinism, American Protestant Thinkers, and the Puzzle of Motivation," in *Darwin's Reception: The Role of Place, Race, Religion, and Gender*, ed. Ronald L. Numbers and John Stenhouse (New York: Cambridge University Press, 1999).

47. Chauncey Wright, "The Genesis of Species," *North American Review* 113 (1871): 63–103, quotation on p. 65; Bowler, *The Non-Darwinian Revolution*, p. 47 and title page.

48. Charles Schuchert and Clara Mae LeVene, *O. C. Marsh: Pioneer in Paleontology* (New Haven: Yale University Press, 1940), pp. 230–234; Thomas H. Huxley, *Science and Culture*, p. 317, quoted ibid., p. 232 (demonstrable fact); Thomas H. Huxley, *American Addresses, with a Lecture on the Study of Biology* (New York: D. Appleton, 1893; first published in 1877), p. 90 (Copernican theory); Charles Darwin to O. C. Marsh, August 31, 1880, quoted in *The Life and Letters of Charles Darwin*, 2:417. For suggestions that natural selection paved the way for the acceptance of evolution, see, e.g., George Daniels, ed., *Darwinism Comes to America* (Waltham, MA: Blaisdell Publishing Co., 1968), pp. xii–xiii, 75; R. J. Wilson, ed., *Darwinism and the American Intellectual: A Book of Readings* (Homewood, IL: Dorsey Press, 1967), pp. 6–7; and Ronald L. Numbers, *Creation by Natural Law: Laplace's Nebular Hypothesis in American Thought* (Seattle: University of Washington Press, 1977), p. 105. In fairness to Wilson, I should point out that in the second edition of *Darwinism and the American Intellectual* (1989), p. 7, he revised his earlier statement.

49. Frank J. Sulloway, *Born to Rebel: Birth Order, Family Dynamics, and Creative Lives* (New York: Pantheon Books, 1996), quotation on p. 237. Sulloway employs a ten-variable model in predicting attitudes toward Darwinism. The three best predictors are birth order, social attitudes, and age, followed by conflict with parents, early parental loss, world travel, national differences, parents' social attitudes, and personal contact with Darwin before 1860. Scientific eminence, which correlates significantly with age, is a significant predictor of opposition to Darwinism. For a brief introduction to Sulloway's approach, see his article "Birth Order, Creativity, and Achievement," in *Encyclopedia of Creativity*, ed. Mark A. Runco and Steven Pritzker (San Diego: Academic Press, in press). Sulloway directs his critique of Marxist explanations of the Darwinian revolution to Adrian Desmond, James Moore, and Robert M. Young.

50. O. C. Marsh, "History and Methods of Paleontological Discovery," *American Journal of Science* 118 (1879): 323–359, quotation on p. 351; Gray, *Darwiniana*, p. 196, from an article, "The Attitude of Working Naturalists toward Darwinism," first published in 1873; E. D. Cope, *The Origin of the Fittest: Essays on Evolution* (New York: D. Appleton, 1887), p. 3, from an article, "Evolution and Its Consequences," first published in 1872. For additional opinions on the appeal of Darwin's theory, see, e.g., Theodore Gill, "The Doctrine of Darwin," *Proceedings of the Biological Society of Washington* 1 (1880–1882): 47–55, especially pp. 50–51; and Joseph LeConte, "The Theory of Evolution and Social Progress," *Monist* 5 (1895): 481–500, especially pp. 485–486.

51. Howard E. Gruber, *Darwin on Man: A Psychological Study of Scientific Creativity,* together with *Darwin's Early and Unpublished Notebooks,* transcribed and annotated by Paul H. Barrett (New York: E. P. Dutton, 1974), pp. 417–418; Gray, *Natural Science and Religion,* pp. 61–62. For a similar statement by the American entomologist Benjamin D. Walsh, see Sorensen, *Brethren of the Net,* p. 200. In *Science and Religion: Some Historical Perspectives* (Cambridge: Cambridge University Press, 1991), p. 271, John Hedley Brooke notes that the commitment to naturalism "had become so much a habit that the belief in a supernatural manifestation or immediate divine agency would automatically be attributed to ignorance or imposture."

52. Theodore Gill, "The Doctrine of Darwin," *Proceedings of the Biological Society of Washington* 1 (1880–81): 47–55, quotation on p. 52; Darwin, *On the Origin of Species,* p. 483. For similar observations by a Harvard anatomist in the Academy, see J[effries] W[yman], Review of *Monograph of the Aye-Aye,* by Richard Owen, *American Journal of Science* 86 (1863): 194–199, quotation on p. 296. T. H. Huxley was making the same points in England; see Adrian Desmond, *Huxley: From Devil's Disciple to Evolution's High Priest* (Reading, MA: Addison-Wesley, 1997), pp. 225–226, 256. Louis Agassiz refused to describe creation in scientific terms; see Jeffries Wyman to Burt G. Wilder, May [?], 1871, quoted in Burt G. Wilder, "Jeffries Wyman, Anatomist (1814–1874)," in *Leading American Men of Science,* ed. David Starr Jordan (New York: Henry Holt, 1910), pp. 171–209, quotation on p. 194; and Nathaniel Southgate Shaler, *The Autobiography of Nathaniel Southgate Shaler* (Boston: Houghton Mifflin, 1909), p. 128.

53. S. R. Calthrop, "Religion and Evolution," *Religious Magazine and Monthly Review* 50 (1873): 193–227, quotation on p. 205; [W. N. Rice], "The Darwinian Theory of the Origin of Species," *New Englander* 26 (1867): 603–635, quotation on p. 608.

54. Gray, *Darwiniana,* pp. 78–79, from a review published in 1860; George F. Wright, "Recent Works Bearing on the Relation of Science to Religion: No. II," *Bibliotheca Sacra* 33 (1876): 448–493, quotation on p. 480. For similar statements from physical scientists in the Academy, see F. A. P. Barnard, "The First Century of the Republic," *Harper's New Monthly Magazine* 52 (1876): 209–233, especially p. 223; and Simon Newcomb, "Address," *Proceedings of the American Association for the Advancement of Science* 27 (1878): 1–28, especially pp. 20–21.

55. See, e.g., A. S. Packard, Jr., "The Law of Evolution," *Independent* 32 (February 5, 1880): 10; Gill, "The Doctrine of Darwin," p. 51; and George R. Agassiz, ed., *Letters and Recollections of Alexander Agassiz with a Sketch of His Life and Work* (Boston: Houghton Mifflin, 1913), p. 116. As a young man, Darwin expressed similar views; see Gruber, *Darwin on Man,* p. 337. I first suggested this interpretation in *Creation by Natural Law,* pp. 108–109. Neal C. Gillespie has made the same point in *Charles Darwin and the Problem of Creation,* where he asserts on p. 147 that "it was more Darwin's insistence on totally natural explanations than on natural selection that won . . . adherence." David L. Hull has also identified Darwin's naturalism as "a major factor in certain scientists' acceptance of evolutionary theory"—as well as in others' rejection of it; see Hull, *Darwin and His Critics: The Reception of Darwin's Theory of Evolution by the Scientific Community* (Cambridge: Harvard University Press, 1973), p. 454.

2. Creating Creationism

This chapter is based on a paper prepared for a conference called "The Evangelical Engagement with Science," sponsored by the Institute for the Study of American Evangelicals and held at Wheaton College, April 1, 1995; and it will appear in the published proceedings, *Evangelicals and Science in Historical Perspective,* ed. David N. Livingstone, D. G. Hart, and Mark A. Noll (New York: Oxford University Press, 1998). I am indebted to the editors of this volume, as well as to Edward J. Larson, Paul A. Nelson, and especially Jon Topham, for their helpful criticisms and suggestions.

1. James Moore, "Deconstructing Darwinism: The Politics of Evolution in the 1860s," *Journal of the History of Biology* 24 (1991): 353–408, quotation on p. 359; Mark A. Noll and David N. Livingstone, "Charles Hodge and the Definition of 'Darwinism,' " in *What Is Darwinism? And Other Writings on Science and Religion,* by Charles Hodge (Grand Rapids, MI: Baker Books, 1994), p. 34.

2. Ronald L. Numbers, *The Creationists* (New York: Alfred A. Knopf, 1992).

3. The *Oxford English Dictionary,* 2nd ed., 20 vols. (Oxford: Clarendon Press, 1989), 3:1135; Charles Hodge, *Systematic Theology,* 3 vols. (New York: Charles Scribner's Sons, 1872), 2:70–71; Charles Darwin to T. H. Huxley, November 25, 1859, *The Correspondence of Charles Darwin,* 10 vols. to date (Cambridge: Cambridge University Press, 1985–), 7:398; Asa Gray, *Darwiniana: Essays and Reviews Pertaining to Darwinism,* ed. A. Hunter Dupree (Cambridge: Harvard University Press, 1963), p. 204, from "The Attitude of Working Naturalists toward Darwinism," originally published in the *Nation,* October 16, 1873; Asa Gray, *Natural Science and Religion: Two Lectures Delivered to the Theological School of Yale College* (New York: Charles Scribner's Sons, 1880), p. 89; J. A. Zahm, *Evolution and Dogma* (Chicago: D. H. McBride, 1896), pp. 73–75; J. W. Dawson, *The Story of the Earth and Man* (New York: Harper & Brothers, 1873), p. 352; [Alexander Agassiz], "A Natural Theory of Creation," *Nation* 8 (1869): 193–194, quotation on p. 193. E. D. Cope contrasted the "creativist doctrine" with the "derivatist doctrine"; Cope, "A Review of the Modern Doctrine of Evolution," *American Naturalist* 14 (1880): 166–179, quotations on p. 166. Darwin also referred to "creationist" or "creationists" in his unpublished essay of 1842, Charles Darwin, *The Foundations of the Origin of Species: Two Essays Written in 1842 and 1844,* ed. Francis Darwin (Cambridge: Cambridge University Press, 1909), pp. 31, 49–50; in his unpublished essay of 1844, ibid., pp. 182, 193; in an 1854 (?) note, *Correspondence,* 5:234; and in two letters to J. D. Hooker in 1856, ibid., 6:170, 304. I am indebted to Jon Topham for information about the Darwin manuscripts and to Sarah Pfatteicher for searching nineteenth-century dictionaries.

4. Charles Darwin, *On the Origin of Species,* with an introduction by Ernst Mayr (Cambridge: Harvard University Press, 1966), p. 437; Tore Frängsmyr, "Linnaeus as a Geologist," in *Linnaeus: The Man and His Work,* ed. Tore Frängsmyr (Berkeley and Los Angeles: University of California Press, 1983), pp. 110–155, especially p. 122; Charles Lyell, *Principles of Geology,* 3 vols. (London: John Murray, 1830–1833), 2:123–126; Louis Agassiz, *Essay on Classification,* ed. Edward Lurie (Cambridge: Harvard University Press, 1962), pp. 173–175; Gray, *Natural Science and*

Religion, p. 35. For Darwin's early views on special creation, see, e.g., his unpublished 1842 and 1844 essays in Darwin, *The Foundations of the Origins of Species*, pp. 22–23, 31, 49–50, 168–171, 182, 191–194, 249–251. Without clarification, he refers to "the common view" of creation in a letter to J. D. Hooker, March 13, 1846, *Correspondence*, 3:300; and in a letter to J. D. Dana, November 11, 1859 (ibid., 7:368). Regarding Agassiz's views, see also Ernst Mayr, "Agassiz, Darwin, and Evolution," *Harvard Library Bulletin* 13 (1959): 165–194; and Mary P. Winsor, "Louis Agassiz and the Species Question," *Studies in History of Biology* 3 (1979): 89–117. For a survey of creationist views before Darwin, see Edwin Tenney Brewster, *Creation: A History of Non-Evolutionary Theories* (Indianapolis: Bobbs-Merrill, 1927), which is the unacknowledged source for much of Frank L. Marsh's chapter "Creationist Theories" in his book *Studies in Creationism* (Washington: Review and Herald Publishing Association, 1950), pp. 22–40.

5. William B. Hayden, *Science and Revelation; or, The Bearing of Modern Scientific Developments upon the Interpretation of the First Eleven Chapters of Genesis* (Boston: Otis Clapp, 1852), p. 77. On pre-Darwinian views of Genesis, see chap. 8, "The Mosaic Story of Creation," in Ronald L. Numbers, *Creation by Natural Law: Laplace's Nebular Hypothesis in American Thought* (Seattle: University of Washington Press, 1977), pp. 88–104.

6. Arnold Guyot, *Creation; or, The Biblical Cosmogony in the Light of Modern Science* (New York: Charles Scribner's Sons, 1884), pp. 116–128; Dawson, *The Story of the Earth and Man*, pp. 340–341, 352; J. W. Dawson, The *Origin of the World, According to Revelation and Science* (Montreal: Dawson Brothers, 1877), pp. 238, 371. On Protestant antievolutionism in the late nineteenth century, see Jon H. Roberts, *Darwinism and the Divine in America: Protestant Intellectuals and Organic Evolution, 1859–1900* (Madison: University of Wisconsin Press, 1988), pp. 209–231.

7. J[effries] W[yman], Review of *Monograph of the Aye-Aye*, by Richard Owen, *American Journal of Science* 86 (1863): 194–199, quotation on p. 196. For a similar observation, see Theodore Gill, "The Doctrine of Darwin," *Proceedings of the Biological Society of Washington* 1 (1880–81): 47–55, quotation on p. 52. On Agassiz's refusal to describe creation, see Jeffries Wyman to Burt G. Wilder, May [?] 1871, quoted in Burt G. Wilder, "Jeffries Wyman, Anatomist (1814–1874)," in *Leading American Men of Science*, ed. David Starr Jordan (New York: Henry Holt, 1910), pp. 171–209, quotation on p. 194; and Nathaniel Southgate Shaler, *The Autobiography of Nathaniel Southgate Shaler* (Boston: Houghton Mifflin, 1909), p. 128. On Gosse, see Frederic R. Ross, "Philip Gosse's *Omphalos*, Edmund Gosse's *Father and Son*, and Darwin's Theory of Natural Selection," *Isis* 68 (1977): 85–96.

8. Numbers, *The Creationists*, p. 49.

9. George McCready Price, *The Phantom of Organic Evolution* (New York: Fleming H. Revell, 1924), pp. 99–100 (burlesque and one act); Geo. E. McCready Price, *Outlines of Modern Christianity and Modern Science* (Oakland, CA: Pacific Press, 1902), pp. 125–127 (libel and dodging). For more on the views of Price, see Numbers, *The Creationists*, pp. 72–101.

10. W. W. Prescott to G. M. Price, November 6, 1908, and C. C. Lewis to G. M. Price, October 23, 1908, both in the Price Papers, Adventist Heritage Center, An-

drews University, Berrien Springs, Michigan; Harold W. Clark, *Back to Creationism* (Angwin, CA: Pacific Union College Press, 1929), p. 135.

11. D. J. Whitney to G. M. Price, December 11, 1935, Price Papers; G. M. Price to B. C. Nelson, July 8, 1935, Byron C. Nelson Papers, Institute for Creation Research, El Cajon, California. On the Religion and Science Association, see Numbers, *The Creationists,* pp. 102–117, from which this account is extracted.

12. D. J. Whitney, "For the Consideration of the Directors of the Religion and Science Association," August 6, 1935, Nelson Papers (obstacle); D. J. Whitney to G. M. Price, December 11, 1935, Price Papers (dogma); D. J. Whitney to G. M. Price, September 9 and 14, 1934, Price Papers (Wheaton). For extant issues of *The Creationist,* see Ronald L. Numbers, ed., *Early Creationist Journals,* vol. 9 of *Creationism in Twentieth-Century America: A Ten-Volume Anthology of Documents, 1903–1961* (New York: Garland Publishing, 1995).

13. Interview with J. Laurence Kulp, July 23, 1984 (pseudo-science); James O. Buswell III, "A Creationist Interpretation of Prehistoric Man," in *Evolution and Christian Thought Today,* ed. Russell L. Mixter (Grand Rapids, MI: Wm. B. Eerdmans, 1959), pp. 165–189, quotations on pp. 169, 188–89. On evolution in the ASA, see Numbers, *The Creationists,* pp. 159–181; and Mark A. Kalthoff, ed., *Creation and Evolution in the Early American Scientific Affiliation,* vol. 10 of *Creationism in Twentieth-Century America.*

14. Russell L. Mixter, *Creation and Evolution* (Wheaton, IL: American Scientific Affiliation, 1950), pp. 1–2; V. R. Edman to the Board of Trustees, Wheaton College, October 28, 1957, Russell L. Mixter Papers, Special Collections, Buswell Library, Wheaton College. On the controversies over evolution at Wheaton College, see Numbers, *The Creationists,* pp. 181–183.

15. Bernard Ramm, *The Christian View of Science and Scripture* (Grand Rapids, MI: Wm. B. Eerdmans, 1954), pp. 180, 228, 293. The phrase "narrow bibliolatry" appears on p. 9 of the paperback edition, also published by Eerdmans. On "progressive creation," see also Edwin K. Gedney, "Geology and the Bible," in *Modern Science and Christian Faith: A Symposium on the Relationship of the Bible to Modern Science,* by Members of the American Scientific Affiliation (Wheaton, IL: Van Kampen Press, 1948), pp. 49–55.

16. Henry M. Morris, *The Troubled Waters of Evolution* (San Diego: Creation-Life Publishers, 1974), p. 16. On the renaissance of flood geology and its rechristening as "creation science," see Numbers, *The Creationists,* pp. 184–257.

17. J. R. Howitt to A. C. Custance, October 22, 1962, December 24, 1965, and August 25, 1973, A. C. Custance Papers, Special Collections, Redeemer College, Ancaster, Ontario, Canada; J. R. Howitt to W. E. Lammerts, [late 1963], Walter E. Lammerts Papers, Bancroft Library, University of California, Berkeley. On creationism in Great Britain, see Numbers, *The Creationists,* pp. 323–330.

18. *Science and Creationism: A View from the National Academy of Sciences* (Washington, D.C.: National Academy Press, 1984), p. 7; Davis A. Young, *Christianity and the Age of the Earth* (Grand Rapids, MI: Zondervan, 1982), p. 10; Howard J. Van Till, Davis A. Young, and Clarence Menninga, *Science Held Hostage: What's Wrong with Creation Science and Evolutionism* (Downers Grove, IL: InterVarsity Press, 1988), pp. 45, 125. In Howard J. Van Till and others, *Portraits of Creation:*

Biblical and Scientific Perspectives on the World's Formation (Grand Rapids, MI: William B. Eerdmans, 1990), written under the auspices of the Calvin Center for Christian Scholarship, Robert E. Snow defines a *"creationist"* as "someone who advocates the development of a science consistent with the view that creation was completed in six twenty-four-hour days, that the Earth and the universe are young (approximately 10,000 years in age), and that most of the Earth's geological features are attributable to the action of a worldwide flood" (p. 167). Two of the best recent studies of creationism in America are Edward J. Larson, *Trial and Error: The American Controversy over Creation and Evolution,* updated ed. (New York: Oxford University Press, 1989); and Christopher P. Toumey, *God's Own Scientists: Creationists in a Secular World* (New Brunswick, NJ: Rutgers University Press, 1994).

3. Darwinism in the American South

This chapter, written with Lester D. Stephens, is based on a paper first presented at a conference entitled "Responding to Darwin: New Perspectives on the Darwinian Revolution," held in Dunedin, New Zealand, May 12–15, 1994; and it will appear in the published proceedings of that conference, *Darwin's Reception: The Role of Place, Race, Religion, and Gender,* ed. Ronald L. Numbers and John Stenhouse (New York: Cambridge University Press, 1999).

1. Monroe Lee Billington, *The American South: A Brief History* (New York: Charles Scribner's Sons, 1971), pp. 301–302; W. J. Cash, *The Mind of the South* (New York: Alfred A. Knopf, 1941), pp. 140–141. For other negative views by historians of the South, see William B. Hesseltine, *A History of the South, 1607–1936* (New York: Prentice-Hall, 1936), p. 340; Clement Eaton, *Freedom of Thought in the Old South* (Durham, NC: Duke University Press, 1940), pp. 312–314; Thomas D. Clark, *The Emerging South,* 2nd ed. (New York: Oxford University Press, 1968), pp. 248–252; John Samuel Ezell, *The South since 1865,* 2nd ed. (New York: Macmillan, 1975), pp. 348–352; and Carl N. Degler, *Place over Time: The Continuity of Southern Distinctiveness* (Baton Rouge: Louisiana State University Press, 1977), p. 23. C. Vann Woodward's classic *Origins of the New South, 1872–1913* (Baton Rouge: Louisiana State University Press, 1951) remains surprisingly silent about evolution in the New South.

2. George M. Marsden, *Understanding Fundamentalism and Evangelicalism* (Grand Rapids, MI: William B. Eerdmans, 1991), pp. 168–173; David N. Livingstone, *Darwin's Forgotten Defenders: The Encounter between Evangelical Theology and Evolutionary Thought* (Grand Rapids, MI: William B. Eerdmans, 1987), p. 124.

3. Jon H. Roberts, *Darwinism and the Divine in America: Protestant Intellectuals and Organic Evolution, 1859–1900* (Madison: University of Wisconsin Press, 1988), p. 222.; A. T. R[obertson], "Darwinism in the South," *Wake Forest Student* 4 (1885): 205–206. James Moore brought this article to my attention.

4. S. E. Morison, *The Oxford History of the United States, 1783–1917,* 2 vols. (London: Oxford University Press, 1927), 2:24; Ronald L. Numbers and Janet S. Numbers, "Science in the Old South: A Reappraisal," in *Science and Medicine in the Old South,* ed. Ronald L. Numbers and Todd L. Savitt (Baton Rouge: Louisiana State University Press, 1988), pp. 9–35.

5. On southern responses to the nebular hypothesis, see Ronald L. Numbers, *Creation by Natural Law: Laplace's Nebular Hypothesis in American Thought* (Seattle: University of Washington Press, 1977), pp. 37–38, 63–64, 86.

6. A. D. Bache, "Remarks upon the Meeting of the American Association at Charleston, S.C., March 1850," in American Association for the Advancement of Science, *Proceedings, Fourth Meeting . . . 1850* (Washington, D.C., 1851); Robert L. Dabney, "Geology and the Bible," *Southern Presbyterian Review* 14 (1861): 246–274; Robert L. Dabney, "A Caution against Anti-Christian Science," in *Discussions by Robert L. Dabney, D.D., LL.D.*, ed. C. R. Vaughan, 4 vols. (Richmond, VA: Presbyterian Committee of Publication, 1892), 3:116–136. On the AAAS meeting, see also Sally Gregory Kohlstedt, *The Formation of the American Scientific Community: The American Association for the Advancement of Science, 1848–60* (Urbana: University of Illinois Press, 1976), p. 116.

7. Michael M. Tuomey, *Report on the Geology of South Carolina* (Columbia, SC: A. S. Johnston, 1848), pp. 58–59; Robert W. Gibbes, *The Present Earth the Remains of a Former World: A Lecture Delivered before the South Carolina Institute, September 6, 1849* (Columbia, SC: A. S. Johnston, 1849), p. 31; R. T. B[rumby], "The Pre-Adamite Earth: Relations of Geology to Theology," *Southern Quarterly Review* 19 (1852): 420–455; [R. T. Brumby], "Relations of Science to the Bible," *Southern Presbyterian Review* 25 (1874): 1–31; [R. T. Brumby], "Gradualness Characteristic of All God's Operations," ibid., pp. 524–555, quotation on p. 540.

8. [James A. Lyon], "The New Theological Professorship—Natural Science in Connexion with Revealed Religion," *Southern Presbyterian Review* 12 (1859): 181–195; E. Brooks Holifield, "Science and Theology in the Old South," in *Science and Medicine in the Old South*, ed. Numbers and Savitt, pp. 127–143, quotation on p. 142. See also Holifield, *The Gentlemen Theologians: American Theology in Southern Culture, 1795–1860* (Durham, NC: Duke University Press, 1978).

9. J. William Flinn, "James Woodrow, A.M., D.D., M.D., LL.D.," in the South Carolina newspaper *Columbia State*, January 18, 1907, pp. 10–11. This biographical account, as well as many pamphlets related to the Woodrow affair, can be found in the John William Flinn Collection, Department of History, Presbyterian Church (U.S.A.), Montreat, North Carolina. The reference to 4,000 constituents appears in "The Trial of Professor Woodrow," *Southern Presbyterian*, September 9, 1886, p. 2.

10. James Woodrow, Editorial Note, *Southern Presbyterian*, May 28, 1885, p. 2.

11. James Woodrow, *Evolution: An Address Delivered May 7th, 1884, before the Alumni Association of the Columbia Theological Seminary* (Columbia, SC: Presbyterian Publishing House, 1884), pp. 17–18, 23, 29. A more colorful version of Woodrow's depiction of evolution from fish to man appeared as a direct quotation from Woodrow in a speech by William Adams, reprinted in "The General Assembly," *Southern Presbyterian*, May 31, 1888, p. 2. For criticism of Woodrow's views on the creation of Eve, see George D. Armstrong, *The Two Books of Nature and Revelation Collated* (New York: Funk & Wagnalls, 1886), p. 94. On Woodrow as a biblical inerrantist, see T. Watson Street, "The Evolution Controversy in the Southern Presbyterian Church with Attention to the Theological and Ecclesiastical Issues Raised," *Journal of the Presbyterian Historical Society* 37 (1959): 234. On Woodrow's endorsement of Guyot, see "Professor Woodrow's Speech before the Synod of

South Carolina," *Southern Presbyterian Review* 36 (1885): 55. For Guyot's views, see Numbers, *Creation by Natural Law,* pp. 91–100.

12. "The Seminary Board Question before the Synod," *Southern Presbyterian,* July 30, 1885, p. 2; John L. Girardeau, *The Substance of Two Speeches on the Teaching of Evolution in Columbia Theological Seminary, Delivered in the Synod of South Carolina, at Greenville, S.C., Oct. 1884* (Columbia, SC: William Sloan, 1885), p. 35; Ernest Trice Thompson, *Presbyterians in the South,* 3 vols. (Richmond, VA: John Knox Press, 1973), 2:464. Thompson offers perhaps the best account of the Woodrow affair, but see also Street, "The Evolution Controversy in the Southern Presbyterian Church," and the largely derivative discussion in Frank Joseph Smith, "The Philosophy of Science in Later Nineteenth Century Southern Presbyterianism," Ph.D. diss., City University of New York, 1992. On Girardeau, see George A. Blackburn, *The Life Work of John L. Girardeau, D.D., LL.D.* (Columbia, SC: The State Company, 1916). On October 24, 1884, the *Greenville Daily News* credited Girardeau with firing "the first shot in the evolution controversy."

13. J. William Flinn, "Evolution and Theology: The Consensus of Science against Dr. Woodrow's Opponents," *Southern Presbyterian Review* 36 (1885): 510; "The Seminary Board Question before the Synod," p. 2; "Columbia Theological Seminary," *Southern Presbyterian,* December 18, 1884, p. 2; "Professor Woodrow's Removal," ibid.

14. "A Sure Enough Subject for the Charleston 'Inquisition,' " *Southern Presbyterian,* November 8, 1888, p. 2; George D. Armstrong, *A Defence of the "Deliverance" on Evolution, Adopted by the General Assembly of the Presbyterian Church in the United States, May 26th, 1886* (Norfolk, VA: John D. Ghiselin, 1886), pp. 3–5.

15. "The Outcome," *Southern Presbyterian,* July 5, 1888, p. 2; "The General Assembly," ibid., May 31, 1888, pp. 1–3. On Woodrow's identification with Galileo, see "Professor Woodrow's Speech before the Synod of South Carolina," pp. 56–58. On his alleged secularity, see "The Perkins Professor's Case," *Southern Presbyterian,* September 10, 1885, p. 2; and *The Examination of the Rev. James Woodrow, D.D., by the Charleston Presbytery* (Charleston, SC: Lucas & Richardson, 1890), p. 1. The quotation about being "routed" appears in Thompson, *Presbyterians in the South,* 2:489. The reference to Draper appears in J. William Flinn, "Evolution and Theology: The Logic of Prof. Woodrow's Opponents Examined," *Southern Presbyterian Review* 36 (1885): 268–304, quotation on p. 270.

16. Flinn, "Evolution and Theology: The Consensus," p. 508; J. Leighton Wilson, "The Evolution Difficulty," *Southern Presbyterian,* September 17, 1885, p. 2.

17. John B. Adger, "The Synod at Cheraw," *Southern Presbyterian,* November 4, 1886, p. 2; "Evolution in the Church," ibid., October 16, 1884, p. 2.

18. Armstrong, *The Two Books,* pp. 86, 96; "The General Assembly," *Southern Presbyterian,* May 27, 1886, p. 2. On Armstrong's standing, see Thompson, *Presbyterians in the South,* 2:477.

19. "Drs. Patton and Hodge on Evolution and the Scriptures," *Southern Presbyterian,* May 6, 1886, p. 2. See also "Sir William Dawson on the Relations of Evolution to the Bible," ibid., May 13, 1886, p. 2.

20. Marsden, *Understanding Fundamentalism and Evangelicalism,* pp. 168–173; Flinn, "Evolution and Theology: The Consensus," p. 579; Emma M. Barnett to

James Woodrow, September 5, 1884, quoted in Smith, "The Philosophy of Science," p. 316.

21. "The Ecclesiastical Blunder," *Southern Presbyterian*, December 11, 1884, p. 2; "More Work for the 'Inquisition,'" ibid., November 29, 1888, p. 2; "Evolution in the South," *New York Times*, April 5, 1885, reprinted in the *New Orleans Daily Picayune*, April 14, 1885, from a copy in James Woodrow's scrapbook, Department of History, Presbyterian Church (U.S.A.), Montreat, North Carolina; Flinn, "Evolution and Theology: The Consensus," pp. 578–579 (emphasis in the original).

22. Leonard Alberstadt, "Alexander Winchell's Preadamites—A Case for Dismissal from Vanderbilt University," *Earth Sciences History* 13 (1994): 97–112; Edwin Mims, *History of Vanderbilt University* (Nashville: Vanderbilt University Press, 1946), pp. 100–101; Paul Conkin, *Gone with the Ivy: A Biography of Vanderbilt University* (Knoxville: University of Tennessee Press, 1985), pp. 50–51, 60–63; Livingstone, *Darwin's Forgotten Defenders*, pp. 86, 91. On pre-Adamism, see David Livingstone, *The Preadamite Theory and the Marriage of Science and Religion*, vol. 82, part 3, of *Transactions of the American Philosophical Society* (1992).

23. Conkin, *Gone with the Ivy*, pp. 63, 97; "Religion and Science at Vanderbilt," *Popular Science Monthly* 13 (1878): 492–495; "Vanderbilt University Again," ibid. 14 (1878): 237–239.

24. William B. Gatewood, Jr., *Preachers, Pedagogues and Politicians: The Evolution Controversy in North Carolina, 1920–1927* (Chapel Hill: University of North Carolina Press, 1966), p. 154; Cash, *Mind of the South*, p. 339.

25. Lester D. Stephens, "Darwin's Disciple in Georgia: Henry Clay White, 1875–1927," *Georgia Historical Quarterly* 78 (1994): 66–91.

26. Ibid.

27. Flinn, "Evolution and Theology: The Consensus," p. 545; "Professor Caldwell at Tulane University," *Southern Presbyterian*, July 23, 1885, p. 2; "What Is It?" ibid., January 19, 1888, p. 2. Flinn pastored a Presbyterian church in New Orleans from 1878 to 1888.

28. John B. Elliott, "President's Address before the New Orleans Academy of Sciences," *Papers Read before the New Orleans Academy of Sciences*, 1886–87, pp. 5–18; Elliott, "The Deeper Revelations of Science: Annual Address before the Academy of Sciences, Feb. 7, 1888," ibid., 1887–88, pp. 398–419.

29. George Little, *Memoirs of George Little* (Tuscaloosa, AL: Weatherford Printing Co., [1924]), p. 101. See also Flinn, "Evolution and Theology: The Consensus," p. 544.

30. On evolution at the University of Virginia and the University of North Carolina, see "Professor Woodrow's Speech before the Synod of South Carolina," pp. 34–35. For the University of Alabama, see James B. Sellers, *History of the University of Alabama* (Tuscaloosa: University of Alabama Press, 1953), pp. 540–541. Woodrow taught evolution in his geology classes at the University of South Carolina. The standard histories of other state universities in the South say virtually nothing about the teaching of evolution.

31. Daniel Walker Hollis, *University of South Carolina*, vol. 2: *College to University* (Columbia: University of South Carolina Press, 1956), pp. 165–180, 245–246.

32. "Professor Woodrow's Speech before the Synod of South Carolina," p. 34; "What Is It?" p. 2; "Inaccurate Reports," *Southern Presbyterian,* November 20, 1884, p. 2; Flinn, "Evolution and Theology: The Consensus," pp. 545–546.

33. "Professor Caldwell at Tulane University," p. 2; "The Southwestern Presbyterian University and Evolution," *Southern Presbyterian,* December 4, 1884, p. 2; Flinn, "Evolution and Theology: The Consensus," pp. 544–545.

34. This paragraph is taken in large part from Ronald L. Numbers, *The Creationists* (New York: Alfred A. Knopf, 1992), p. 40.

35. Regarding Wofford, see "Professor Woodrow's Speech before the Synod of South Carolina," p. 34. Regarding Guilford, see Joseph Moore, "The Greatest Factor in Human Evolution," *Guilford Collegian* 6 (1894): 240–244; T. Gilbert Pearson, "Evolution in Its Relation to Man," ibid. 8 (1896): 107–111; Pearson, *Adventures in Bird Protection: An Autobiography* (New York: D. Appleton-Century, 1937), pp. 58–59; and Oliver H. Orr, Jr., *Saving American Birds: T. Gilbert Pearson and the Founding of the Audubon Movement* (Gainesville: University Press of Florida, 1992), pp. 19, 42, 46, 49, 81.

36. J. Lawrence Smith, "Address," *Proceedings of the American Association for the Advancement of Science* (Portland, ME, 1873), pp. 14–16. On medical opinion, see, e.g., J. C., Review of *On the Origin of Species by Means of Natural Selection,* by Charles Darwin, *Richmond and Louisville Medical Journal* 9 (1870): 84–100; and F. M. Robertson, "President's Address," *Transactions of the South Carolina Medical Association,* 1880–81, Appendix, pp. 1–15. I am indebted to the late Patricia Spain Ward for these last two references.

37. John McCrady, "The Law of Development by Specialization: A Sketch of Its Probable Universality," *Journal of the Elliott Society of Natural History* 1 (1860): 101–114; Lester D. Stephens and Dale R. Calder, "John McCrady of South Carolina: Pioneer Student of North American Hydrozoa," *Archives of Natural History* 19 (February 1992): 39–54; Numbers, *The Creationists,* pp. 8–9.

38. John Bachman, "An Investigation of the Cases of Hybridity in Animals, Considered in Reference to the Unity of the Human Species," *Charleston Medical Journal* 5 (1850): 168–197, especially p. 186; Lester D. Stephens, *Ancient Animals and Other Wondrous Things: The Story of Francis Simmons Holmes, Paleontologist and Curator of the Charleston Museum* (Charleston, SC: Charleston Museum, 1988); Lewis R. Gibbes, marginalia on letter from John L. Girardeau to Gibbes, March 5, 1891, in the Lewis R. Gibbes Papers, Manuscript Division, Library of Congress. See also Lester D. Stephens, "Lewis R. Gibbes and the Professionalization of Science in Antebellum South Carolina," unpublished paper presented at the annual meeting of the Southern Historical Association, Atlanta, November 13, 1980; and Stephens, "A Sketch of Natural History Collecting in Charleston, South Carolina: The Golden Age, 1820–1865," unpublished paper presented at the second North American meeting of the Society for the History of Natural History, Pittsburgh, October 24, 1986.

39. Tamara Miner Haygood, "Henry Ravenel (1814–1887): Views on Evolution in Social Context," *Journal of the History of Biology* 21 (1988): 457–472; Gabriel E. Manigault, manuscript autobiography written ca. 1887–1897, in the Manigault Family Papers, Manuscripts Department, Library, University of North Carolina

at Chapel Hill; J. H. Mellichamp to George Englemann, August 26, 1872, in the South Carolina Collection, Charleston Museum Library.

40. Silas McDowell, undated (ca. 1865 or later) manuscript entitled "Evolution" in the Silas McDowell Papers, Manuscripts Department, Library, University of North Carolina at Chapel Hill; Edmund Berkeley and Dorothy Smith Berkeley, *A Yankee Botanist in the Carolinas: The Reverend Moses Ashley Curtis, D.D. (1808–1872)* (Berlin: J. Cramer, 1986).

41. "The Evolution Hypothesis," *Southern Review* 3 (1868): 408–440, quotation on p. 419. For somewhat more critical assessments in the same journal, see "The Origin of Species," ibid. 9 (1871): 700–728; "Darwinism," ibid. 12 (1873): 406–423; and "Philosophy versus Darwinism," ibid. 13 (1873): 253–273.

42. [W. S. Bean], "The Outlook of Modern Science," *Southern Presbyterian Review* 25 (1874): 331–338, quotation on p. 335. The Robertson quotation appears in James Moore, The *Darwin Legend* (Grand Rapids, MI: Baker Books, 1994), p. 119.

43. Numbers, *The Creationists,* pp. 46–49.

44. Ibid., p. 41; "Fighting Evolution at the Fundamentals Convention," *Christian Fundamentals in School and Church* 7 (July/September 1925): 5.

45. Maynard Shipley, *The War on Modern Science: A Short History of the Fundamentalist Attacks on Evolution and Modernism* (New York: Alfred A. Knopf, 1927), pp. 75–186, quotation on p. 141. See also Norman F. Furniss, *The Fundamentalist Controversy, 1918–1931* (New Haven: Yale University Press, 1954), pp. 76–100.

46. F. D. Perkins, "Evolution Theory Taught in American Schools," *Western Recorder,* August 11, 1921, p. 4; J. Frank Norris to William Jennings Bryan, December 28, 1923, Box 38, Bryan Papers, Manuscript Division, Library of Congress; Numbers, *The Creationists,* p. 47.

4. The Scopes Trial

This chapter grew out of a lecture given at a symposium entitled "Religion and Public Life: Seventy Years after the Scopes Trial," organized by the Robert Penn Warren Center for the Humanities, Vanderbilt University, November 2, 1995. I am grateful to John Cejka, Sarah Pfatteicher, and Jennifer Munger for their research assistance; to Paul Boyer for sharing his collection of American history textbooks; and to Paul K. Conkin, Edward J. Larson, Michael Lienesch, Jon H. Roberts, and John Stenhouse for their encouragement and suggestions. Edward J. Larson's excellent study of the Scopes trial, *Summer for the Gods: The Scopes Trial and America's Continuing Debate over Science and Religion* (New York: Basic Books, 1997), appeared after I had written this chapter, as did Carol Iannone's insightful essay, "The Truth about *Inherit the Wind,*" *First Things* 70 (February 1997): 28–33.

1. William E. Ellis, "Scopes Trial," in *Encyclopedia of Southern Culture,* ed. Charles Reagan Wilson and William Ferris (Chapel Hill: University of North Carolina Press, 1989), pp. 1376–1377; *National Standards for United States History: Exploring the American Experience* (Los Angeles: National Center for History in the Schools, 1995), p. 180; "Church Meets State," *U.S. News & World Report* 118 (April 24, 1995): 28. See, e.g., the following textbooks from the early 1990s: Gary B. Nash, *American Odyssey: The United States in the Twentieth Century* (Lake Forest,

IL: Glencoe, 1991), pp. 307–308; John A. Garraty, *The Story of America* (Austin, TX: Holt, Rinehart and Winston, 1991), pp. 868–869; Winthrop D. Jordan and others, *The Americans: A History* (Evanston, IL: McDougal, Littell, 1991), pp. 614–615; Thomas V. DiBacco, *History of the United States* (Boston: Houghton Mifflin, 1991), p. 528; Joseph R. Conlin, *Our Land, Our Time: A History of the United States*, 2nd ed. (Austin, TX: Holt, Rinehart and Winston, 1991), p. 653. Although the majority of textbooks that I examined discussed the Scopes trial, Warren A. Nord found no mention of the trial in "the five most commonly used texts in North Carolina"; William A. Nord, *Religion and American Education: Rethinking a National Dilemma* (Chapel Hill: University of North Carolina Press, 1995), pp. 288–289.

2. Martin E. Marty, *Righteous Empire: The Protestant Experience in America* (New York: Dial Press, 1970), p. 220; George M. Marsden, *Understanding Fundamentalism and Evangelicalism* (Grand Rapids, MI: William B. Eerdmans, 1991), p. 60; Robert D. Linder, "Fifty Years after Scopes: Lessons to Learn, a Heritage to Reclaim," *Christianity Today* 19 (1975): 1009–1012, quotation on p. 1010; Raymond A. Eve and Francis B. Harrold, *The Creationist Movement in Modern America* (Boston: Twayne Publishers, 1991), pp. 25–26; George E. Webb, *The Evolution Controversy in America* (Lexington: University Press of Kentucky, 1994), p. 93. Among the best accounts of the trial and its aftermath are Ferenc M. Szasz, "The Scopes Trial in Perspective," *Tennessee Historical Quarterly* 30 (1971): 288–298; Paul M. Waggoner, "The Historiography of the Scopes Trial: A Critical Re-Evaluation," *Trinity Journal*, n.s., 5 (1984): 155–174; and Edward J. Larson, *Trial and Error: The American Controversy over Creation and Evolution*, updated ed. (New York: Oxford University Press, 1989), pp. 58–92. On the Scopes trial as "Fundamentalism's last stand," see, e.g., William Warren Sweet, *The Story of Religion in America* (New York: Harper & Brothers, 1939), p. 572, where the quotation appears; Richard Hofstadter, *Anti-Intellectualism in American Life* (New York: Alfred A. Knopf, 1969), p. 130; and Dorothy Nelkin, *The Creation Controversy: Science or Scripture in the Schools* (New York: W. W. Norton, 1982), p. 32. On the trial as "really the beginning"—or at least not the end—of the antievolution crusade, see, e.g., Szasz, "The Scopes Trial in Perspective," p. 289, where the quotation appears; Edwin Scott Gaustad, *A Religious History of America*, rev. ed. (San Francisco: Harper & Row, 1990), p. 263; and William Vance Trollinger, Jr., *God's Empire: William Bell Riley and Midwestern Fundamentalism* (Madison: University of Wisconsin Press, 1990), p. 50.

3. Ferenc Morton Szasz, "Three Fundamentalist Leaders: The Roles of William Bell Riley, John Roach Straton, and William Jennings Bryan in the Fundamentalist-Modernist Controversy," Ph.D. diss., University of Rochester, 1969, p. 351; Ernest R. Sandeen, *The Roots of Fundamentalism: British and American Millenarianism, 1800–1930* (Chicago: University of Chicago Press, 1970), p. 163; Virginia Gray, "Anti-Evolution Sentiment and Behavior: The Case of Arkansas," *Journal of American History* 57 (1970): 352–366, quotation on p. 359 (emphasis added). On urban support for the Tennessee antievolution bill, see Kenneth K. Bailey, "The Enactment of Tennessee's Antievolution Law," *Journal of Southern History* 16 (1950): 472–510, especially pp. 488–489.

4. Ray Ginger, *Six Days or Forever? Tennessee v. John Thomas Scopes* (Boston: Beacon Press, 1958); L. Sprague de Camp, *The Great Monkey Trial* (Garden City,

NY: Doubleday, 1968); Ronald L. Numbers, *The Creationists* (New York: Alfred A. Knopf, 1992), p. 41.

5. See, e.g., John T. Scopes and James Presley, *Center of the Storm: Memoirs of John T. Scopes* (New York: Holt, Rinehart and Winston, 1967), written by the defendant; and Arthur Garfield Hays, *Let Freedom Ring* (New York: Boni and Liveright, 1928), written by one of the defense attorneys.

6. For a stenographic record of the trial, see *The World's Most Famous Court Trial: Tennessee Evolution Case* (Cincinnati: National Book Co., 1925).

7. The *Nashville Tennessean,* quoted in "The End of the Scopes Case," *Literary Digest* 92 (February 5, 1927): 14–15, quotation on p. 15; Howard W. Odum, "The Duel to the Death," *Social Forces* 4 (1925): 189–194, quotations on p. 190; "Science Brings Scopes Trial to Midwest's Ears," *Chicago Daily Tribune,* July 15, 1925, p. 3; *The World's Most Famous Court Trial.*

8. "Ended at Last," *New York Times,* July 22, 1925, p. 18; "Darrow and Bryan Shed Court Courtesy for Post-Trial Opinions of One Another," *Nashville Tennessean,* July 22, 1925, p. 8.

9. Stephen Jay Gould, *Hen's Teeth and Horse's Toes* (New York: W. W. Norton, 1983), pp. 272–273. In *Let Freedom Ring,* pp. 73, 75, Arthur Garfield Hays charges Bryan with believing both that "the world was created in 4004 B.C." and that creation might have taken "millions of years." In *Defender of the Faith: William Jennings Bryan: The Last Decade, 1915–1925* (New York: Oxford University Press, 1965), p. 350, Lawrence W. Levine charges Bryan with committing "a serious blunder" in revealing his views of Genesis.

10. *The World's Most Famous Court Trial,* pp. 296–303.

11. Numbers, *The Creationists,* pp. x–xiii, 17–19, 43–46, and passim.

12. *The World's Most Famous Court Trial,* p. 285; Numbers, *The Creationists,* pp. 26–27, 45, 66–67; W. B. Riley and Harry Rimmer, *A Debate: Resolved, That the Creative Days in Genesis Were Aeons, Not Solar Days* (n.p., [1929]), reprinted in Ronald L. Numbers, ed., *Creation-Evolution Debates,* vol. 2 of *Creationism in Twentieth-Century America: A Ten-Volume Anthology of Documents, 1903–1961,* ed. Ronald L. Numbers (New York: Garland Publishing, 1995), pp. 393–425.

13. William Jennings Bryan, Letter to the Editor, *Forum* 70 (1923): 1852–53, which was kindly brought to my attention by Paul M. Waggoner. On the role of miracles in the creation, see W. J. Bryan to S. J. Bole, July 27, 1922, S. James Bole Papers, Nebraska State Historical Society.

14. W. J. Bryan to Howard A. Kelly, June 22, 1925, Box 47, W. J. Bryan Papers, Library of Congress.

15. Numbers, *The Creationists,* pp. 72–101; "Letter to the Editor from the Principal Scientific Authority of the Fundamentalists," *Science* 63 (1926): 259. On Bryan's alleged betrayal of Fundamentalism, see, e.g., William E. Leuchtenburg, *The Perils of Prosperity, 1914–32* (Chicago: University of Chicago Press, 1958), pp. 222–223; Gerraty, *The Story of America,* p. 869; and Gould, *Hen's Teeth and Horse's Toes,* pp. 272–273.

16. G. M. Price to W. J. Bryan, July 1, 1925, Box 47, Bryan Papers; "Says Millions Here Oppose Darwinism," *New York Times,* September 8, 1925, p. 9; George McCready Price, "What Christians Believe about Creation," *Bulletin of*

Deluge Geology 2 (1942): 65–84; G. M. Price to Cyril B. Courville, May 24, 1942, G. M. Price Papers, Adventist Heritage Center, Andrews University, Berrien Springs, Michigan. See also George McCready Price, "The Scopes Trial—1925," *These Times* 69 (February 1960): 4–7.

17. James J. Thompson, Jr., *Tried as by Fire: Southern Baptists and the Religious Controversies of the 1920s* (Macon, GA: Mercer University Press, 1982), pp. 131–132 (which quote Norris); W. B. Riley, "Bryan the Great Commoner and Christian," *Christian Fundamentals in School and Church* 7 (October/December, 1925): 5–11, 37, quotation on p. 11; W. B. Riley, "The World's Christian Fundamentals Association and the Scopes Trial," ibid., pp. 37–48, quotation on p. 39. On Riley, see also Trollinger, *God's Empire,* p. 50.

18. George F. Milton, "A Dayton Postscript," *Outlook* 140 (1925): 550–553, quotation on p. 551; "Topics of the Times," *New York Times,* July 16, 1925, p. 18 (regarding prayer); Dixon Merritt, "The Theatrical Performance at Dayton," *Outlook* 140 (1925): 390–392 (regarding chancery court); David E. Lilienthal, "The Tennessee Case and State Autonomy," ibid., pp. 453–454 (regarding national norms); "The Conduct of the Scopes Trial," *New Republic* 43 (1925): 331–333, quotations on p. 332; Walter Lippmann, *American Inquisitors: A Commentary on Dayton and Chicago* (New York: Macmillan, 1928), pp. 12–14. For additional criticism of Bryan's attackers, see "Tennessee *vs.* Truth," *Nation* 121 (1925): 58; and George Henry Payne, "Speaking As an Episcopalian," *Forum* 74 (1925): 425–429. See also Ginger, *Six Days or Forever,* p. 174; and Norman Furniss, *The Fundamentalist Controversy, 1918–1931* (New Haven: Yale University Press, 1954), p. 91.

19. "Changing Sentiment," *Nashville Tennessean,* July 21, 1925, p. 4 (quoting the *Boston Post*); de Camp, *The Great Monkey Trial,* pp. 444–448; "Says Dayton Trial Was Only Skirmish," *New York Times,* August 11, 1925, p. 8 (quoting Neal); "Amateur Dramatics at Dayton," *Christian Century* 42 (1925): 969–970.

20. Hofstadter, *Anti-Intellectualism in American Life,* p. 130. For a contrasting view, see, e.g., de Camp, *The Great Monkey Trial,* p. 473. The newspapers I examined are the *Nashville Tennessean, Chicago Tribune, New York Times, Washington Post,* and *San Francisco Chronicle.*

21. "Malone Talks at Follies," *New York Times,* July 25, 1925, p. 13; Philip Kinsley, "Bryan to Lead Holy War into 7 Legislatures," *Chicago Sunday Tribune,* July 19, 1925, p. 4; Waggoner, "The Historiography of the Scopes Trial," pp. 161–164; Frederick Lewis Allen, *Only Yesterday: An Informal History of the Nineteen-Twenties* (New York: Harper & Brothers, 1931), p. 206. On Allen's immense influence, see Roderick Nash, *The Nervous Generation: American Thought, 1917–1930* (Chicago: Rand McNally, 1970), pp. 5–6.

22. This and the following two paragraphs closely follow my retrospective review of *Inherit the Wind* in *Isis* 84 (1993): 763–764. See also Jerome Lawrence and Robert E. Lee, *Inherit the Wind* (New York: Bantam Books, 1985; first published in 1955).

23. Scopes and Presley, *Center of the Storm,* pp. 33–34, 53, 59–62, 270.

24. Ibid., p. 270; Lawrence and Lee, *Inherit the Wind,* p. 86; *National Standards for United States History,* p. 180.

25. "W. J. Bryan Dies in His Sleep at Dayton," *New York Times,* July 27, 1925, p. 1; "Commoner and Darrow Tilt in War of Words," *Chicago Sunday Tribune,* July 19, 1925, p. 1; "Bryan Maps Campaign in Legislatures," *San Francisco Chronicle,* July 19, 1925, p. 1; "Straton to Debate Darrow, He Hopes," *New York Times,* August 4, 1925, p. 10; Hubert C. Herring, "Please, Mr. Potter! Please, Mr. Straton!" *Christian Century* 42 (1925): 1215–1216, quotation on p. 1215; Maynard Shipley, *The War on Modern Science: A Short History of the Fundamentalist Attacks on Evolution and Modernism* (New York: Alfred A. Knopf, 1927), p. 48, quoting Clarke. See also Walter White's comment in "Scopes Guilty by Consent of Counsel," *San Francisco Chronicle,* July 22, 1925, p. 7; and Kinsley, "Bryan to Lead Holy War," p. 4.

26. Jerry R. Tompkins, ed., *D-Days at Dayton: Reflections on the Scopes Trial* (Baton Rouge: Louisiana State University Press, 1965), pp. 50–51, quoting Mencken; "Darrow and Bryan Shed Court Courtesy," pp. 1, 8, quoting Malone; "Says Dayton Trial Was Only Skirmish," p. 8, quoting Neal; "Lays Scopes Trial to Publicity Thirst," *New York Times,* August 5, 1925, p. 10, quoting Potter; "Dayton—And After," *Nation* 121 (1925): 155–156.

27. Miriam Allen de Ford, "After Dayton: A Fundamentalist Survey," *Nation* 122 (1926): 604–605, quotation on p. 604; "Religion and Science in Tennessee," *Round Table* 15 (1925): 732–748, quotation on p. 740; Willard B. Gatewood, Jr., *Preachers, Pedagogues and Politicians: The Evolution Controversy in North Carolina, 1920–1927* (Chapel Hill: University of North Carolina Press, 1966), p. 162. On Bryan's death as a spur to the antievolution campaign, see also Ginger, *Six Days or Forever?* p. 211. For predictions that Bryan's death would send the campaign into decline, see, e.g., Watson Davis and Frank Thone, "Science and Intellectual Freedom," *Nature* 116 (1925): 284.

28. "Campaigning for Genesis," *Literary Digest* 92 (February 19, 1927): 33, quoting Dabney and Herbert S. Hadley of Washington University; Maynard Shipley, "Growth of the Anti-Evolution Movement," *Current History* 32 (1930): 330–332, quotation on p. 330.

29. Shipley, "Growth of the Anti-Evolution Movement," p. 330; Orland Kay Armstrong, "Bootleg Science in Tennessee," *North American Review* 227 (1929): 138–142; de Ford, "After Dayton," p. 604; Judith V. Grabiner and Peter D. Miller, "Effects of the Scopes Trial," *Science* 185 (1974): 832–837. For contrary views of the effects of the trial, see George Gaylord Simpson, Letter to the Editor, *Science* 187 (1975): 389; and Philip J. Pauly, "The Development of High School Biology: New York City, 1900–1925," *Isis* 82 (1991): 662–688, especially pp. 663–664, 685–687.

30. de Camp, *The Great Monkey Trial,* p. 477 (boredom); Larson, *Trial and Error,* p. 83. On Fundamentalism after 1925, see Joel A. Carpenter, *Revive Us Again: The Reawakening of American Fundamentalism* (New York: Oxford University Press, 1997); and Numbers, *The Creationists,* pp. 102–103.

31. Henry M. Morris, *History of Modern Creationism* (San Diego: Master Book, 1984), pp. 65–69; Jerry Falwell, with Ed Dobson and Ed Hinson, ed., *The Fundamentalist Phenomenon: The Resurgence of Conservative Christianity* (Garden City, NY: Doubleday, 1981), pp. 85–86, 89. In *Redeeming America: Piety and Politics in the New Christian Right* (Chapel Hill: University of North Carolina Press, 1993), p. 154,

Michael Lienesch, one of the few scholars to pay any attention to conservative images of the trial, attributes this view to the Christian right.

32. Harry Rimmer, *The Theories of Evolution and the Facts of Science* (Los Angeles: Research Science Bureau, 1929), pp. 4–6; Scott M. Huse, *The Collapse of Evolution* (Grand Rapids, MI: Baker Book House, 1983), p. 98. See also Roger Daniel Schultz, "All Things Made New: The Evolving Fundamentalism of Harry Rimmer, 1890–1952," Ph.D. diss., University of Arkansas, 1989, pp. 183–185.

33. John Wolf and James S. Mellett, "The Role of 'Nebraska Man' in the Creation-Evolution Debate," *Creation/Evolution* 5, no. 2 (1985): 31–43. Nevertheless, Osborn continued to cite Nebraska Man after the trial; see Henry Fairfield Osborn, *Evolution and Religion in Education: Polemics of the Fundamentalist Controversy of 1922–1926* (New York: Charles Scribner's Sons, 1926), pp. 102–106.

34. Bolton Davidheiser, *Evolution and Christian Faith* ([Philadelphia]: Presbyterian and Reformed Publishing Co., 1969), pp. 88–103, quotations on pp. 97–98. Davidheiser's essay on the trial was apparently reprinted in a work I have not seen: "The Scopes Trial," in *A Symposium on Creation III*, ed. D. W. Patten (Grand Rapids, MI: Baker Book House, 1971). For evidence of Davidheiser's influence, see, e.g., Robert E. Kofahl, *Handy Dandy Evolution Refuter*, rev. ed. (San Diego: Beta Books, 1980), pp. 106–108; and Bert Thompson, *The History of Evolutionary Thought* (Fort Worth: Star Bible & Tract Corp., 1981), pp. 157–164.

35. Bill Keith, *Scopes II: The Great Debate* (Shreveport, LA: Huntington House, 1982), pp. vii–viii (bigotry); W. R. Bird, *The Origin of Species Revisited: The Theories of Evolution and of Abrupt Appearance*, 2 vols. (New York: Philosophical Library, 1989), 1:8–9, 2:377; Duane T. Gish, *Teaching Creation Science in Public Schools* (El Cajon, CA: Institute for Creation Research, 1995), p. v; *The World's Most Famous Court Trial*, p. 187. Regarding Darrow, Bird cites Adell Thompson, *Biology, Zoology, and Genetics: Evolution Model vs. Creation Model* (1983), p. 271, which I have not seen.

5. "Sciences of Satanic Origin"

An earlier version of this chapter was published in *Spectrum* 9 (January 1979): 17–30, and the current version is printed here with the permission of the editor, Roy Branson. Roughly the first half of the chapter appeared originally as "Science Falsely So-Called: Evolution and Adventists in the Nineteenth Century," *Journal of the American Scientific Affiliation*, 27 (March 1975): 18–23, and was reprinted in *Spectrum* with the permission of the editor, Richard H. Bube.

1. On Ellen W. White and the rise of Adventism, see Ronald L. Numbers, *Prophetess of Health: A Study of Ellen G. White* (New York: Harper & Row, 1976); and Ronald L. Numbers and Jonathan M. Butler, ed., *The Disappointed: Millerism and Millenarianism in the Nineteenth Century* (Bloomington: Indiana University Press, 1987).

2. Ronald L. Numbers, *Creation by Natural Law: Laplace's Nebular Hypothesis in American Thought* (Seattle: University of Washington Press, 1977), pp. 105–118. See also Rodney Lee Stiling, "The Diminishing Deluge: Noah's Flood in Nineteenth-Century American Thought," Ph.D. diss., University of Wisconsin–Madison, 1991.

3. Ellen G. White, "Science and Revelation," *Signs of the Times,* March 13, 1884, p. 161; Ellen G. White, *Spiritual Gifts: Important Facts of Faith, in Connection with the History of Holy Men of Old* (Battle Creek, MI: Seventh-day Adventist Publishing Assn., 1864), p. 93.

4. "Science and the Bible," *Advent Review and Sabbath Herald,* February 24, 1859, p. 107. This statement is attributed to a Dr. Cumming, probably John Cumming, the Scottish divine known for his studies of biblical prophecies. Hereafter the *Advent Review and Sabbath Herald,* popularly known as the *Review and Herald,* will be cited as *R & H.*

5. J. H. Kellogg, *Harmony of Science and the Bible on the Nature of the Soul and the Doctrine of the Resurrection* (Battle Creek, MI: Review and Herald Publishing Assn., 1879), pp. 10–11, 28–29.

6. [John Harvey Kellogg], "Evolution," *Health Reformer* 9 (May 1874): 159. See also "Darwinism," ibid. 8 (December 1973): 381. Later Kellogg seems to have become a theistic evolutionist; see Richard W. Schwarz, *John Harvey Kellogg, M.D.* (Nashville: Southern Publishing Assn., 1970), p. 190.

7. A. T. Jones, "The Uncertainty of Geological Science," *R & H,* August 21, 1883, p. 530; "Geology and the Bible," ibid., October 17, 1865, p. 157; [Uriah Smith], "False Theories of Geologists," ibid., September 5, 1882, p. 568. Smith had attended Phillips Exeter Academy for several years, but financial considerations had prevented him from going on to college.

8. "The Blunders of Geologists," *R & H,* October 24, 1865, pp. 161–162; Jones, "The Uncertainty of Geological Science," p. 529; [Smith], "False Theories of Geologists," p. 568; [G. W. Amadon], "The Skeptic Met," *R & H,* September 4, 1860, p. 121.

9. Robert Patterson, "Geological Chronology," *R & H,* February 8, 1870, p. 51, reprinted, with an editorial introduction, from an article in *Family Treasury;* Joseph F. Tuttle, "That Old Skull," *R & H,* October 25, 1870, p. 146, reprinted from a work by Tuttle.

10. A. T. Jones, "The Uncertainty of Geological Science," *R & H,* August 7, 1883, pp. 497–498; August 14, 1883, pp. 513–514; August 21, 1883, 529–530, quotations on pp. 513, 530. The following year Jones published another series, on " 'Evolution' and Evolution," ibid., March 11, 1884, pp. 162–163; March 18, 1884, pp. 178–179; March 25, 1884, pp. 194–195.

11. J. O. Corliss, "Geologists vs. the Mosaic Record," *R & H,* February 19, 1880, pp. 116–117; Ellen G. White, "Science and the Bible in Education," *Signs of the Times,* March 20, 1884, p. 177; Stephen Pierce, "Does the Bible Agree with Science?" *R & H,* October 3, 1871, p. 121; and Jones, " 'Evolution' and Evolution," p. 195. The quotation regarding the methods of science appeared in [D. N. Lord], "The Structure of the Earth," ibid., February 12, 1880, p. 99. Lord was a non-Adventist known for his writings on science and religion and on the fulfillment of Bible prophecies. For examples of treating supernatural explanations as unscientific, see R. F. Cottrell, "The Antiquity of Man," ibid., January 21, 1875, p. 29; and Jones, "The Uncertainty of Geological Science," p. 530.

12. John Hall, "Turning the Tables," *R & H,* April 9, 1872, p. 130. For similar views expressed by Adventists, see "Too Knowing for Faith," ibid., November 8, 1877, p. 148; and Corliss, "Geologists vs. the Mosaic Record," p. 116.

13. [Lord], "The Structure of the Earth," p. 98; David N. Lord, *Geognosy; or, The Facts and Principles of Geology against Theories,* 2nd ed. (New York: Franklin Knight, 1857); Maurice Hodgen, ed., *School Bells and Gospel Trumpets: A Documentary History of Seventh-day Adventist Education in North America* (Loma Linda, CA: Adventist Heritage Publications, 1978), p. 18. Uriah Smith made an equally strong pronouncement in "Giving Way," *R & H,* October 23, 1883, p. 664.

14. White, *Spiritual Gifts,* p. 90.

15. Ibid., p. 91; [James White], "The First Week of Time," *R & H,* February 12, 1880, pp. 104–105. E. J. Waggoner, an Adventist physician and minister, repeated Mrs. White's views in *The Literal Week,* no. 18 in the Apples of Gold Library (Oakland, CA: Pacific Press Publishing Co., 1894), p. 2.

16. J. P. Henderson, "The Bible—No. 7," *R & H,* July 5, 1887, p. 419; J. N. Andrews, "The Memorial of Creation," ibid., April 7, 1874, p. 129. In 1860 the editors of the *R & H* reprinted a passage from *The Bible True,* by the Presbyterian minister William Plumer, advocating an interpretation similar to Henderson's; see "Geology," *R & H,* July 3, 1860, p. 49. Kellogg in his early years also seems to have leaned toward this view; see *Harmony of Science and the Bible,* p. 20. For views similar to Andrews's, see D. T. Bourdeau, "Geology and the Bible; or, A Pre-Adamic Age of Our World Doubtful," *R & H,* February 5, 1867, p. 98; and "The Creation of Light," ibid., September 28, 1869, p. 112.

17. Numbers, *Creation by Natural Law,* pp. 77–87; P. R. Russel, "Darwinism Examined," *R & H,* May 18, 1876, p. 153; T. DeWitt Talmage, "Evolution: Anti-Bible, Anti-Science, Anti-Commonsense," ibid., April 24, 1883, p. 261. Talmage was a popular Presbyterian minister.

18. "Carlyle on Darwin," *R & H,* January 17, 1878, p. 19; Talmage, "Evolution," p. 261; W. H. Littlejohn, "The Temple in Heaven," *R & H,* February 24, 1885, p. 116. Uriah Smith explained the "defection of leading Christians . . . to the ranks of the evolutionists" as being the result of satanic influence; see "Giving Way," p. 664. On the widespread teaching of evolution in American colleges, see "Do Our Colleges Teach Evolution?" *Independent,* December 18, 1879, pp. 14–15.

19. Russel, "Darwinism Examined," p. 153. See also Adolphus Smith, "Science, Falsely So Called," *R & H,* July 8, 1873, p. 31.

20. White, *Spiritual Gifts,* pp. 92–95. See also Waggoner, *The Literal Week,* p. 3.

21. White, *Spiritual Gifts,* p. 75; [Uriah Smith], *The Visions of Mrs. E. G. White: A Manifestation of Spiritual Gifts According to the Scriptures* (Battle Creek, MI: SDA Publishing Assn., 1868), pp. 102–104. Regarding this work, James White wrote: "I felt very grateful to God that our people could have this able defense of those views which they so much love and prize, and which others despise and oppose. This book is designed for a very wide circulation." James White, "New and Important Work," *R & H,* August 25, 1868, p. 160. See also Gordon Shigley, "Amalgamation of Man and Beast: What Did Ellen White Mean?" *Spectrum* 12, no. 4 (1982): 10–19.

22. [Smith], *The Visions of Mrs. E. G. White,* pp. 74–75. See also Bourdeau, "Geology and the Bible," pp. 98–99.

23. On Price, see Ronald L. Numbers, *The Creationists* (New York: Alfred A. Knopf, 1992), pp. 72–101; and Harold W. Clark, *Crusader for Creation: The Life and Writings of George McCready Price* (Mountain View, CA: Pacific Press, 1966).

24. G. M. Price to H. W. Clark, June 15, 1941, Price Papers, Adventist Heritage Center, Andrews University Library, Berrien Springs, Michigan. See also George McCready Price, "Some Early Experiences with Evolutionary Geology," *Bulletin of Deluge Geology* 1 (November 1941): 77–92.

25. George McCready Price, *Illogical Geology: The Weakest Point in the Evolution Theory* (Los Angeles: Modern Heretic Co., 1906), p. 9; G. M. Price to Martin Gardner, May 13, 1952, courtesy of Martin Gardner.

26. William G. Moorehead to G. M. Price, November 6, 1906; David Starr Jordan to G. M. Price, August 28, 1906; David Starr Jordan to G. M. Price, May 5, 1911; G. M. Price to Harold W. Clark, November 20, 1941; all in the Price Papers.

27. George McCready Price, *The New Geology* (Mountain View, CA: Pacific Press, 1923), pp. 637–638.

28. Charles Schuchert, Review of *The New Geology*, by George McCready Price, *Science*, May 30, 1924, pp. 486–487; George McCready Price, Letter to the Editor, ibid., March 5, 1926, p. 259; G. M. Price to J. McKeen Cattell, March 22, 1926, Price Papers; *Science*, March 5, 1926, p. 259; G. M. Price to Molleurus Couperus, November 1946, courtesy of Molleurus Couperus. On Price's reputation, see also Martin Gardner, *Fads and Fallacies in the Name of Science* (New York: Dover Publications, 1957), p. 127.

29. William Jennings Bryan to G. M. Price, June 7, 1925, Box 47, Bryan Papers, Library of Congress; G. M. Price to William Jennings Bryan, July 1, 1925, ibid.; L. Sprague de Camp, *The Great Monkey Trial* (Garden City, NY: Doubleday, 1968), pp. 401–402. For Price's view of Bryan's performance, see George McCready Price, "What Christians Believe about Creation," *Bulletin of Deluge Geology* 2 (October 1942): 76–77.

30. G. M. Price to E. S. Ballenger, September 28, 1925, in my possession; "Shout Down American in Evolution Debate," *New York Times*, September 7, 1925, p. 5; "Says Millions Here Oppose Darwinism," ibid., September 8, 1925, p. 9. For the actual debate, see *Is Evolution True? Verbatim Report of Debate between George McCready Price, M.A., and Joseph McCabe* (London: Watts, 1925). Two of Price's Adventist disciples, Francis D. Nichol and Alonzo L. Baker, had recently debated the evolutionist Maynard Shipley in San Francisco; see Alonzo L. Baker, "The San Francisco Evolution Debates, June 13–14, 1925," *Adventist Heritage* 2 (Winter 1975): 23–31.

31. Discussion of George McCready Price, "Revelation and Evolution: Can They Be Harmonized?" *Journal of the Transactions of the Victoria Institute* 47 (1925): 183, 187. See also George McCready Price, "Geology and Its Relation to Scripture Revelation," ibid. 46 (1924): 97–123.

32. Harold W. Clark, "What is 'Flood Geology'?" *Naturalist* 26 (Summer 1966): 10–11; interview with Harold W. Clark, May 11, 1973.

33. Harold W. Clark to Richard M. Ritland, June 2, 1963, courtesy of Richard M. Ritland; interview with Clark, May 11, 1973.

34. Harold W. Clark to G. M. Price, April 9, April 23, April 30, August 14, and October 20, 1940; G. M. Price to Harold W. Clark, April 21 and June 9, 1940; all in the Price Papers. See also Glenn Calkins to the Members of the Committee Appointed to Listen to Professor George McCready Price and to Professor H. W.

Clark, April 6, 1941, Publishing Department Papers, General Conference of Seventh-day Adventists. Price unconvincingly insisted that he had filed not a charge of heresy, but rather "a charge of slander, of legal libel"; see G. M. Price to Harold W. Clark, October 23, 1944, Price Papers. Although Price did not mention Clark by name in his booklet, he did name Clark's publications; see George McCready Price, *Theories of Satanic Origin* (Loma Linda, CA; Published by the author, n.d.). A letter that accompanied copies of the pamphlet further identified the target as a professor "from our college at Angwin"; see George McCready Price to Fellow Workers, n.d., courtesy of Molleurus Couperus.

35. L. E. Froom to G. M. Price, July 12, 1942, Price Papers; Harold W. Clark, *The New Diluvialism* (Angwin, CA: Science Publications, 1946). In 1944 Clark wrote: "Unfortunately, most of our attention up to this time has been directed toward the negative aspects of the problem. We have been intent on finding the flaws in the evolution theory, but have neglected to build up a positive creationism which would give one a definite concept of the creation doctrine in relation to the data of the various sciences." Harold W. Clark, "The Positive Aspects of Creationism," *Ministry,* May 1944, p. 5.

36. Harold W. Clark, *Genes and Genesis* (Mountain View, CA: Pacific Press, 1940); Frank Lewis Marsh, *Evolution, Creation, and Science* (Washington: Review and Herald Publishing Assn., 1944). Regarding Marsh, see his "Life Summary of a Creationist," unpublished ms., December 1, 1968, courtesy of the late F. L. Marsh. On the differences between Marsh and Price, see Frank L. Marsh to G. M. Price, September 5, 1943, Price Papers. Despite their disagreements Price praised Marsh's book; see G. M. Price to Harold W. Clark, October 23, 1944, Price Papers.

37. Clark, *Crusader for Creation,* p. 58; "Bulletin Mailing List, etc.—1944, 1945, 1946," courtesy of Molleurus Couperus; G. M. Price to H. W. Clark, October 23, 1944, Price Papers. *The Bulletin of Deluge Geology and Related Sciences* appeared in five volumes between June 1941, and December 1945. In 1946 it was superseded by *The Forum for the Correlation of Science and the Bible,* which continued until 1948 under the editorship of Molleurus Couperus.

38. George McCready Price, "In the Beginning," *The Forum for the Correlation of Science and the Bible* 1 (1946–47): 9; Molleurus Couperus, "Some Remarks Regarding the Radioactive Time Estimation of the Age of the Earth, ibid. 2 (1947–48): 119; Harold W. Clark, "In Defense of the Ultra-Literal View of the Creation of the Earth," ibid. 1 (1946–47): 11; Molleurus Couperus to H. W. Clark, May 8, 1947, courtesy of Molleurus Couperus.

39. Minutes of the Autumn Council, October 25, 1957, 226th Meeting, General Conference Committee, vol. 19, book 5, p. 1004; minutes of the 259th Meeting of the General Conference Committee, April 24, 1958, ibid., p. 1156; minutes of the 108th Meeting of the General Conference Committee, January 21, 1960, vol. 20, book 3, p. 511; all in the Archives of the General Conference of Seventh-day Adventists, Silver Spring, Maryland.

40. Marsh, "Life Summary of a Creationist," p. 5a; Frank L. Marsh to G. M. Price, March 14, 1962, Price Papers.

41. P. Edgar Hare to Richard Hammill, February 7, 1963; R. R. Figuhr to P. Edgar Hare, February 5, 1964; both courtesy of P. Edgar Hare.

42. Interview with Frank Lewis Marsh, August 30, 1972; Marsh, "Life Summary of a Creationist," pp. 5b–5c.

43. Interview with Richard M. Ritland, May 7, 1973; Richard M. Ritland, *A Search for Meaning in Nature: A New Look at Creation and Evolution* (Mountain View, CA: Pacific Press, 1970). For evidence of Ritland's influence, see G. M. Price to F. L. Marsh, January 10, 1960, courtesy of F. L. Marsh; and H. W. Clark to F. L. Marsh and others, November 19, 1960, courtesy of R. M. Ritland.

44. *R & H,* October 10, 1968, p. 23, quoted in Marsh, "Life Summary of a Creationist," p. 5c.

45. "Tentative Creation Statement," *Spectrum* 8 (January 1977): 58.

46. On the Adventist origins of scientific creationism, see Numbers, *The Creationists,* passim.

47. James L. Hayward, "The Many Faces of Adventist Creationism: 1980–1995," *Spectrum* 25 (March 1996): 16–34.

48. Ibid.; Floyd Petersen, "Science Faculty Vary in Views on Creationism," *Adventist Today* 2 (November/December, 1994): 19.

6. Creation, Evolution, and Holy Ghost Religion

An earlier version of this chapter appeared in *Religion and American Culture* 2 (1992): 127–158, and the current version is printed here with the permission of the editors. The chapter was prepared in collaboration with Tim Kruse, who traveled the Midwest in pursuit of pertinent documents and tutored me on the history of Wesleyan theology. The staff and scholars associated with the Wesleyan/Holiness Studies Project—including Paul Bassett, Dan Bays, Edith Blumhofer, David Bundy, Mel Dieter, Mel Robeck, Susie Stanley, and especially Bill Kostlevy—repeatedly helped me bibliographically and conceptually. Dawn Corley searched the pages of the *Pentecostal Herald.* Clyde R. Root, of the Hal Bernard Dixon, Jr., Pentecostal Research Center in Cleveland, Tennessee, and Joyce Lee, of the Assemblies of God Archives in Springfield, Missouri, guided me to useful Pentecostal sources.

1. See, e.g., Melvin Esterday Dieter, *The Holiness Revival of the Nineteenth Century* (Metuchen, NJ: Scarecrow Press, 1980).

2. See, e.g., Donald W. Dayton, *Theological Roots of Pentecostalism* (Grand Rapids, MI: Francis Asbury Press, 1987); and Edith L. Blumhofer, *Restoring the Faith: The Assemblies of God, Pentecostalism, and American Culture* (Urbana: University of Illinois Press, 1993).

3. Andrew Johnson, "The Evolution Articles," *Pentecostal Herald* 38 (September 29, 1926): 6.

4. The exception is David N. Livingstone, *Darwin's Forgotten Defenders: The Encounter between Evangelical Theology and Evolutionary Thought* (Grand Rapids, MI: William B. Eerdmans, 1987). Although not denominationally oriented, Jon H. Roberts, *Darwinism and the Divine in America: Protestant Intellectuals and Organic Evolution, 1859–1900* (Madison: University of Wisconsin Press, 1988), draws on a number of Methodist sources. James R. Moore, *The Post-Darwinian Controversies: A Study of the Protestant Struggle to Come to Terms with Darwin in Great Britain*

and America, 1870–1900 (Cambridge: Cambridge University Press, 1979), mentions Methodists only in passing.

5. See Ronald L. Numbers, "Creationism in Twentieth-Century America," *Science* 218 (1982): 538–544; and Ronald L. Numbers, *The Creationists* (New York: Alfred A. Knopf, 1992).

6. Andrew Johnson, "Evolution Outlawed by Science [No. 3]," *Pentecostal Herald* 37 (December 9, 1925): 5. See also Andrew Johnson, "Reply to Dr. E. L. Powell's Sermon on Evolution," ibid. 34 (March 15, 1922): 6.

7. William North Rice, *Twenty-Five Years of Scientific Progress and Other Essays* (New York: Thomas Y. Crowell, 1894), p. 7 and passim. For a similar observation regarding declining opposition to evolution, by a colleague of Rice's in the biology department at Wesleyan University, see H. W. Conn, *Evolution To-Day* (New York: G. P. Putnam's Sons, 1886), p. 18.

8. E. Brooks Holifield, "The English Methodist Response to Darwin," *Methodist History* 10 (January 1972): 22; Windsor Hall Roberts, "The Reaction of the American Protestant Churches to the Darwinian Philosophy, 1860–1900," Ph.D. diss., University of Chicago, 1936, p. 196. On Methodist reactions to inorganic evolution, see Ronald L. Numbers, *Creation by Natural Law: Laplace's Nebular Hypothesis in American Thought* (Seattle: University of Washington Press, 1977), pp. 120–121.

9. Livingstone, *Darwin's Forgotten Defenders*, pp. 85–91; Rice, *Twenty-Five Years of Scientific Progress*, pp. 52–53.

10. L. T. Townsend, *Collapse of Evolution* (New York: American Bible League, 1905), p. 47; L. T. Townsend, *Evolution or Creation: A Critical Review of the Scientific and Scriptural Theories of Creation and Certain Related Subjects* (New York: Fleming H. Revell, 1896), pp. 133–134, 154; L. T. Townsend, *Adam and Eve: History or Myth?* (Boston: Chapple, 1904), p. 83; A. C. Dixon to G. F. Wright, May 16 and 24, 1910, G. F. Wright Papers, Oberlin College Archives. For biographical detail, see Moore, *Post-Darwinian Controversies*, pp. 198–199.

11. See, e.g., Arthur T. Pierson, *"Many Infallible Proofs": The Evidences of Christianity; or, The Written and Living Word of God* (London: Morgan and Scott, n.d.), pp. 109–126, by a Presbyterian fellow traveler; Randolph S. Foster, *Creation: God in Time and Space*, vol. 4 of *Studies in Theology* (New York: Hunt and Eaton, 1895), pp. 26–36, by a Methodist bishop sympathetic to the Holiness revival; and Luke Woodard, *What Is Truth?* (Auburn, NY: Knapp, Peck and Thomson, 1901), pp. 131–157, by a Quaker with Holiness leanings. See also Luke Woodard, *Evolution: Unscientific and Antibiblical* (n.p., n.d.). On the attitude of Holiness Friends toward evolution, see Thomas D. Hamm, *The Transformation of American Quakerism: Orthodox Friends, 1800–1907* (Bloomington: Indiana University Press, 1988), p. 109. For a critical view of efforts to harmonize Genesis and geology by a Wesleyan Methodist physician, see John Bingham Hunt, *Genesis and Geology: Fact versus Fiction* (Syracuse, NY: Wesleyan Methodist Publishing Assn., 1901), pp. 3–4. The reference to "Holy Ghost religion" comes from W. B. Godbey, *Bible Astronomy* (Louisville: Pentecostal Publishing Co., n.d.), p. 3.

12. Godbey, *Bible Astronomy*, pp. 7–8; W. B. Godbey, *Man the Climax of Creation* (n.p., n.d.), pp. 3–7; W. B. Godbey, *Regenerated Earth* (Greensboro, NC: Apostolic

Messenger Office, n.d.), pp. 3–5. Because Godbey turned eighty before the outbreak of World War I, I assume he wrote these undated pamphlets by then.

13. L. N. Moore, "Original Sin and Evolution," *Christian Witness and Advocate of Bible Holiness* 28 (January 13, 1898): 2–3. On Conklin, see J. W. Atkinson, "E. G. Conklin on Evolution: The Popular Writings of an Embryologist," *Journal of the History of Biology* 18 (1985): 31–50.

14. *Annual Minutes of the Thirty-Eight Conferences of the Free Methodist Church* (Chicago: Free Methodist Publishing House, 1903), p. 79; Adelaide Lionne Beers, *The Romance of a Consecrated Life: A Biography of Alexander Beers* (Chicago: Free Methodist Publishing House, 1922), p. 318.

15. Paul Merritt Bassett, "The Fundamentalist Leavening of the Holiness Movement, 1914–1940: The Church of the Nazarene: A Case Study," *Wesleyan Theological Journal* 13 (1978): 65–91.

16. "The General Holiness Convention," *Pentecostal Herald* 35 (October 17, 1923): 4. Emphasis added.

17. "Wherein the Evolutionists and the Tongues' People Are Alike," *Free Methodist* 68 (August 23, 1935): 4–5. Two Holiness churches, the Church of God (Anderson, Indiana) and the Salvation Army, generally avoided the issue in the 1920s, although the *Handbook of Salvation Army Doctrine* (New York: The Salvation Army, 1923) included a section called "The Origin of Man," pp. 46–49, which affirmed belief in the Genesis account. I am indebted to Roger J. Green (Salvation Army) and Merle D. Strege (Church of God) for information about these denominations. In contrast, the tiny Church of God (Holiness), centered in Missouri and Kansas, repeatedly denounced Darwinism in *The Church Herald and Holiness Banner;* see Clarence Eugene Cowen, "A History of the Church of God (Holiness)," Ph.D. diss., University of Missouri, 1948, pp. 171–173. Membership data for Holiness churches in 1926 are found in Charles Edwin Jones, *Perfectionist Persuasion: The Holiness Movement and American Methodism, 1867–1936* (Metuchen, NJ: Scarecrow Press, 1974), p. 210.

18. E. P. Ellyson, *Is Man an Animal?* (Kansas City, MO: Nazarene Publishing House, 1926); Basil W. Miller and U. E. Harding, *"Cunningly Devised Fables": Modernism Exposed and Refuted* (n.p., [1925]), pp. 17, 26, 129. See also E. P. Ellyson, *Doctrinal Studies* (Kansas City, MO: Nazarene Publishing House, [1936]), pp. 63–68. For the publication history of the Miller and Harding book, see Bassett, "The Fundamentalist Leavening of the Holiness Movement," pp. 77–78.

19. "Rev. B. T. Roberts on Evolution," *Free Methodist* 58 (April 7, 1925): 11; G. M. Price, "Two Startling Fossils That Confound Evolution," *Free Methodist* 59 (April 6, 1926): 5, 10–11. An index to the *Free Methodist* is available at the Historical Center, Free Methodist Headquarters, Winona Lake, Indiana. I am grateful to Fran Haslam for her help in using the records at the Center. Price's biggest Holiness boosters were Arthur K. White and Ray B. White, sons of the founder of the Pillar of Fire movement and authors of *A Toppling Idol—Evolution* (Zarephath, NJ: Pillar of Fire, 1933).

20. "The Controversy Continues," *Wesleyan Methodist* 84 (August 24, 1927): 1; George D. Watson, *God's First Words: Studies in Genesis: Historic, Prophetic and Experimental* (New York: Fleming H. Revell, 1919), p. 17. Advertisements for Bry-

an's books appeared in the May 9, 1928, and January 16, 1929, issues of the *Wesleyan Methodist.* Daniel L. Burnett, Director of Archives, International Center—The Wesleyan Church, Indianapolis, provided assistance in locating Wesleyan Methodist materials.

21. B. H. Shadduck, *Jocko-Homo: The Heaven-Bound King of the Zoo* (Louisville: Pentecostal Publishing Co., n.d.), pp. 1–2; B. H. Shadduck, *Puddle to Paradise* (Rogers, OH: Jocko-Homo Publishing Co., 1925); B. H. Shadduck, *Rastus Augustus Explains Evolution* (Rogers, OH: Jocko-Homo Publishing Co., 1928), p. 2. See also, e.g., J. Thos. Johnson, "Science Falsely So-Called," *Pilgrim Holiness Advocate* 5 (August 20, 1925): 1.

22. Vinson Synan, *The Old-Time Power* (Franklin Springs, GA: Advocate Press, 1973), p. 187; Kenneth Wireman, "A Comparative Study of the Effect of the Teaching of Biology on Student Attitudes toward Organic Evolution in Assemblies of God Church Schools," Ph.D. diss., University of Utah, 1971, pp. 53–54. See also Otto J. Klink, *Why I Am Not an Evolutionist* (Springfield, MO: Gospel Publishing House, 1931), by an Assemblies of God evangelist; and William A. Williams, *The Evolution of Man Scientifically Disproved in Fifty Arguments* (Camden, NJ: William A. Williams, 1928), by an educator whose wife was an Assemblies of God minister. An index to the *Pentecostal Evangel* is available at the Assemblies of God Archives in Springfield, Missouri.

23. Synan, *The Old-Time Power,* pp. 186–187; George Floyd Taylor, "Exposition on Genesis 1–3," ms. in the Emmanuel College Library, Franklin Springs, Georgia. The *Pentecostal Holiness Advocate* published this manuscript in twenty-eight parts, beginning with the vol. 15, January 14, 1932, issue and ending with the vol. 16, July 21, 1932, issue. I am grateful to Jeff Trexler for bringing this document to my attention and to Rachel Howard for making a copy available to me.

24. C. M. Robeck, Jr., "Aimee Semple McPherson," in *Dictionary of Pentecostal and Charismatic Movements,* ed. Stanley M. Burgess and Gary B. McGee (Grand Rapids, MI: Zondervan Publishing House, 1988), p. 568; *There Is a God! Debate between Aimee Semple McPherson, Fundamentalist, and Charles Lee Smith, Atheist* (Los Angeles: Foursquare Publications, [1934]).

25. "Opposition of Science Falsely So-Called," *Church of God Evangel* 10 (June 28, 1919): 1–2.

26. [H. C. Morrison], "The Coming Controversy," *Pentecostal Herald* 33 (October 19, 1921): 1, 8. On Morrison, see Percival A. Wesche, *Henry Clay Morrison: Crusader Saint* (Berne, IN: Herald Press, 1963); and Melvin E. Dieter, "Henry Clay Morrison," in *Encyclopedia of Religion in the South,* ed. Samuel S. Hill (Macon, GA: Mercer University Press, 1984), pp. 514–515. By 1928 the number of antievolution articles in the *Pentecostal Herald* had dropped precipitously, although Morrison included evolution among the themes he planned to emphasize that year; H. C. Morrison, "Suggestive Themes for Discussion in the Herald Next Year," *Pentecostal Herald* 40 (January 4, 1928): 8. On Johnson, see Willard B. Gatewood, Jr., *Preachers, Pedagogues, and Politicians: The Evolution Controversy in North Carolina, 1920–1927* (Chapel Hill: University of North Carolina Press, 1966), p. 192; and an untitled note in the *Pentecostal Herald* 39 (February 2, 1927): 6. Large sections of Johnson's

"Evolution Outlawed by Science" were recycled under the title "Fatal Gaps in Evolution" in the *American Holiness Journal* in 1949–50.

27. On Bryan, see, e.g., Newton Wray, "In Memoriam—William Jennings Bryan," *Pentecostal Herald* 38 (June 2, 1926): 3, 6; and C. F. Wimberly, "Modern Apostles of Faith, Chapter XXVII: William Jennings Bryan," ibid. 39 (September 21, 1927): 3–6. During the early 1920s, Bryan's byline frequently appeared in the *Herald.* On Townsend, see H. C. Morrison, "A Foreword," ibid. 33 (August 24, 1921): 8; H. C. Morrison, "The Collapse of Evolution," ibid. 34 (March 8, 1922): 5; and H. C. Morrison, "The Collapse of Evolution," ibid. 41 (May 15, 1929): 8.

28. H. C. Morrison, "The Collapse of Evolution," *Pentecostal Herald* 41 (May 15, 1929): 8; Chas. E. Robinson, *The Not-Ashamed Club* (Springfield, MO: Gospel Publishing Assn., 1927), pp. 80–85. I am indebted to Mel Robeck for bringing this story to my attention. See also L. L. Pickett, "Evolution: Part III," *Pentecostal Herald* 38 (January 6, 1926): 3.

29. Miller and Harding, *"Cunningly Devised Fables,"* pp. 110–111; E. G. Burritt to C. A. Blanchard, n.d., and J. L. Brasher to C. A. Blanchard, April 23, 1919, both in the Charles A. Blanchard Papers, Wheaton College Archives. Andrew Johnson repeatedly attacked Methodist colleges in his series "Methodist and Modern Thought," which appeared in vol. 33 of the *Pentecostal Herald;* see, e.g., no. 2 (August 3, 1921): 4; no. 3 (August 10, 1921): 5; and no. 7 (September 7, 1921): 4. See also [H. C. Morrison], "Unscientific Teaching," ibid. 34 (August 23, 1922): 1.

30. [H. C. Morrison], "Will It Stand the Scientific Test?" *Pentecostal Herald* 34 (August 2, 1922): 8; Gatewood, *Preachers, Pedagogues, and Politicians,* pp. 79–80; Kenneth K. Bailey, *Southern White Protestantism in the Twentieth Century* (New York: Harper and Row, 1964), 53–54; H. C. Morrison, "An Open Letter to My Dear Bishop," *Pentecostal Herald* 39 (May 11, 1927): 1, 8.

31. Paul Merritt Bassett, "Culture and Concupiscence: The Changing Definition of Sanctity in the American Holiness Movement, 1867–1920," unpublished paper presented to the Scholars Seminar of the Wesleyan/Holiness Study Project, Shakertown, Kentucky, June 9, 1990; "Attention! Carolina Bible Training School," *Bible Student* 1 (April/June, 1926): 1. See also C. W. B[ridwell], "Colorado Methodism and Counterfeit Holiness," *Rocky Mountain Pillar of Fire* 6 (February 15, 1905): 6–7, and the accompanying cartoon on p. 1.

32. Andrew Johnson, "Methodism and Modern Thought [no. 9]," *Pentecostal Herald* 33 (October 12, 1921): 7; H. C. Morrison, "Evolutionists in Trouble," ibid. 35 (August 15, 1923): 8–9; [H. C. Morrison], "The Anti-Darwin Bill," ibid. 34 (March 8, 1922): 1.

33. Ellyson, *Is Man an Animal?*, p. 62; G. W. Ridout, *Dr. Fosdick Answered: An Exposé of Christian Liberalism* (Louisville: Pentecostal Publishing Co., n.d.), p. 9; H. C. Morrison, "Destructive Criticism and the Second Coming of Christ: A Series of Open Letters to Dr. Geo. P. Mains: Fourteenth Letter," *Pentecostal Herald* 33 (March 9, 1921): 9; Andrew Johnson, "Methodism and Modern Thought [no. 1] ibid. 33 (July 20, 1921): 4; Len G. Broughton, "The New Way," ibid. 38 (July 21, 1926): 4.

34. H. C. Morrison, "An Open Letter to a Young Preacher [no. 9]," *Pentecostal Herald* 38 (October 27, 1926): 1, 8; L. L. Pickett, *God or the Guessers: Some Strictures on Present Day Infidelity* (Louisville: Pentecostal Publishing Co., 1926), p. 11.

35. Andrew Johnson, "Evolution Outlawed by Science [no. 23]," *Pentecostal Herald* 38 (May 26, 1926): 4; Andrew Johnson and L. L. Pickett, *Postmillennialism and the Higher Critics* (Chicago: Glad Tidings Publishing Co., n.d.), p. 50. On Johnson's knowledge of biology, see the letter from H. G. Baker, assistant professor of biology at Southwestern University, to Johnson, reprinted in "Evolution Outlawed by Science [no. 2]," *Pentecostal Herald* 37 (December 2, 1925): 3.

36. Andrew Johnson, "Evolution Outlawed by Science [no. 26]," *Pentecostal Herald* 38 (June 30, 1926): 4; H. C. Morrison, "Destructive Criticism and the Second Coming . . . Twenty-First Letter," ibid. 33 (May 11, 1921): 8; A. R. Higgs, "The Age of the Universe," ibid. 40 (March 28, 1928): 5; Andrew Johnson, "Evolution Outlawed by Science [no. 5]," ibid. 37 (December 23, 1925): 4. See also L. L. Pickett, "In the Beginning—God—Gen. 1:1," ibid. 38 (April 7, 1926): 5.

37. S. A. Steel, "The Menace of Rationalism," *Pentecostal Herald* 33 (November 2, 1921): 9; Samuel A. Steel, *The Modern Theory of the Bible* (New York: Fleming H. Revell, 1921), pp. 6–7, 75, 84, 87–89, 96. See also S. A. Steel, "The Holiness Movement," *Revivalist* 6 (May 1893): 1. For biographical information, see "Samuel Augustus Steel (1849–1934)," in *Encyclopedia of World Methodism*, ed. Nolan B. Harmon (Nashville: United Methodist Publishing House, 1974), pp. 2241–2243. For a typical reference to LeConte, see Andrew Johnson, "Evolution Outlawed by Science [no. 3]," *Pentecostal Herald* 37 (December 9, 1925): 9.

38. Broughton, "The New Way," pp. 4–5.

39. Townsend, *Collapse of Evolution*, pp. 47–53; W. C. Curtis, "Three Letters Bearing upon the Controversy over Evolution," *Science* 61 (1925): 648; Albert Fleischmann, "The Doctrine of Organic Evolution in the Light of Modern Research," *Journal of the Transactions of the Victoria Institute* 65 (1933): 194–214.

40. Pickett, *God or the Guessers*, p. 19; Andrew Johnson, "Evolution Outlawed by Science [no. 7]," *Pentecostal Herald* 38 (January 13, 1926): 4; Andrew Johnson, "Evolution Outlawed by Science [no. 26]," ibid. 38 (June 30, 1926): 9. For a typical list, see C. F. Wimberly, "Evolution—Ten Reasons Why I Do Not Believe It," ibid. 37 (November 4, 1924): 4.

41. S. J. Bole, *The Modern Triangle: Evolution, Philosophy and Criticism* (Los Angeles: Bible Institute of Los Angeles, 1926), p. 13; George W. Ridout, "Two Notable Books," *Pentecostal Herald* 38 (July 28, 1926): 39. For biographical information, I have relied primarily on S. J. Bole, *Confessions of a College Professor* (Los Angeles: Biola Book Room, 1922); p. 43; Franklin W. Scott, ed., *The Semi-Centennial Alumni Record of the University of Illinois* (Champaign-Urbana: University of Illinois, 1918), p. 749; and Bole's personnel file at Wheaton College. See also S. J. Bole, *The Battlefield of Faith* (University Park, IA: College Press, 1940).

42. G. M. Price to E. S. Ballenger, April 20, 1927, and January 30, 1928, both courtesy of Donald Mote; D. J. Whitney to L. A. Higley, August 13, 1935, and D. J. Whitney to Byron C. Nelson, August 26, 1935, both in the Nelson Papers, Institute for Creation Research, El Cajon, California. On Higley, see *National Cyclopedia of American Biography*, 43:363.

43. Molleurus Couperus to W. J. Tinkle, April 19, 1944, courtesy of Molleurus Couperus; W. J. Tinkle to G. M. Price, January 4, 1940, Price Papers, Adventist

Heritage Collection, James White Library, Andrews University, Berrien Springs, Michigan.

44. F. Alton Everest, "The American Scientific Affiliation: Its Growth and Early Development," unpublished ms. in the ASA Collection, Wheaton College Archives, especially Appendix 26, which duplicates the program for the second annual convention; William J. Tinkle and Walter E. Lammerts, "Biology and Creation," in *Modern Science and Christian Faith,* by Members of the American Scientific Affiliation, 3rd printing (Wheaton, IL: Van Kampen Press, 1950), pp. 58–97. Denominational affiliations are given in the *Journal of the American Scientific Affiliation* 3 (September 1951): vii. In 1960 the ASA met at Seattle Pacific College; in 1961 at Houghton College; in 1963 at Asbury College; and in 1974 at Bethany Nazarene College. I am indebted to Mark Kalthoff for lending me a copy of the Everest manuscript.

45. On the history of the Creation Research Society, see Numbers, *The Creationists,* pp. 214–240. Members of the steering committee are identified in the first issue of the *Creation Research Society Quarterly* 1 (July 1964): [13].

46. William Sims Bainbridge and Rodney Stark, "Superstitions: Old and New," *Skeptical Inquirer* 4 (Summer 1980): 20; James F. Gregory, Review of *The Genesis Flood,* by John C. Whitcomb, Jr., and Henry M. Morris, *Free Methodist* 94 (October 17, 1961): 14; S. Hugh Paine, Review of *The Genesis Flood,* by John C. Whitcomb, Jr., and Henry M. Morris, *Wesleyan Methodist* 119 (June 7, 1961): 14; S. Hugh Paine, Jr., "In the Beginning, God Created," *Houghton Milieu* 54 (March 1979): 2–7; S. Hugh Paine to Ronald L. Numbers, November 10, 1989. After the appearance of *The Genesis Flood,* Morris gave an invited lecture at Houghton College; Henry M. Morris, *A History of Modern Creationism* (San Diego: Master Books, 1984), p. 161. In the 1963 survey reported by Bainbridge and Stock, only 57 percent of Church of God members expressed opposition to human evolution.

47. Cecil B. Hamann and J. Paul Ray, "Progressive Creationism," *Good News* 15 (March/April 1982): 12–14; Cecil B. Hamann, videotaped "Lectures on Paleontology and Early Man," 1982, in the Asbury College Archives; interviews with John Brushaber, James Behnke, and William Toll, March 20, 1989. Information about early faculty and courses comes from the *Asbury College Bulletins,* found in the Asbury College Archives. Hamann gave *The Genesis Flood* a surprisingly positive review in *Recent Books: A Quarterly Review for Ministers* 3 (October/December 1961): 45. In his autobiography, *Ten College Generations* (New York: American Press, 1956), Jay B. Kenyon does not mention having once taught chemistry and biology. I am indebted to Ivan Zabilka for information about Hamann.

48. Morris, *A History of Modern Creationism,* p. 259; interview with Henry M. Morris and Duane Gish, October 26, 1980; Turner Collins and Gary Liddle, unpublished syllabus for a "Seminar in Science and Religion" offered at Evangel College, spring 1979; L. Duane Thurman, *How to Think about Evolution and Other Bible-Science Controversies* (Downers Grove, IL: InterVarsity Press, 1978); Myrtle M. Fleming, "Evolution: Do We Know What We Are Talking About?" unpublished paper read at the Society for Pentecostal Studies, Des Moines, Iowa, November 5, 1971, and deposited in the Hal Bernard Dixon, Jr., Pentecostal Research Center, Lee College, Cleveland, Tennessee. On Fleming, see Vinson Synan, *Emmanuel*

College: The First Fifty Years, 1919–1969 (Franklin Springs, GA: Emmanuel College Library, 1969), p. 112. See also the published proceedings of a 1977 Bible-Science Symposium sponsored by Lee College: Robert H. O'Bannon, Lois Beach, and J. Patrick Daugherty, ed., *Science and Theology: Friends or Foes?* (Cleveland, TN: Pathway Press, 1977).

49. Lee Haines, "The Book of Genesis," in *The Wesleyan Bible Commentary*, 6 vols. (Grand Rapids, MI: William B. Eerdmans, 1967), 1:21–28.

50. George Herbert Livingston, "Genesis," in *Beacon Bible Commentary*, 10 vols. (Kansas City, MO: Beacon Hill Press, 1969), 1:32–33, 56–60. See also H. Orton Wiley, *Christian Theology*, vol. 1 (Kansas City, MO: Kingshighway Press, 1940), pp. 455–468.

51. Eugene F. Carpenter, "Cosmology," in *A Contemporary Wesleyan Theology: Biblical, Systematic, and Practical*, 2 vols. (Grand Rapids, MI: Francis Asbury Press, 1983), 1:177–178. In the chapter "Anthropology" in *Contemporary Wesleyan Theology*, 1:196–198, Charles W. Carter defended the special creation of humans. Carpenter's criticism of Wiley incorrectly gives the impression that the Nazarene theologian chose the gap theory over other interpretations of Genesis, but in fact he wrote equally positively about the day-age theory; see Wiley, *Christian Theology*, 1:465.

52. John Wesley, *Explanatory Notes upon the Old Testament*, 3 vols. (Bristol: William Pine, 1765), 1:6, 9; John Wesley, "Remarks on the Count de Buffon's 'Natural History,'" in *The Works of John Wesley*, 14 vols. (Grand Rapids, MI: Zondervan, 1958), 13:452; John Wesley, "God's Approbation of His Works," ibid., 4:213; Alexander Hardie, *Evolution: Is It Philosophical, Scientific, or Scriptural?* (Los Angeles: Times-Mirror Press, 1924), p. 222; Warren A. Candler, "John Wesley and Evolution," *Pentecostal Herald* 37 (November 25, 1925): 4. On Wesley as a proto-evolutionist, see Frank W. Collier, *John Wesley among the Scientists* (New York: Abingdon Press, 1928), pp. 170–174. See also Adam Clarke, *The Holy Bible, Containing the Old and New Testaments . . . with a Commentary and Critical Notes* (New York: Methodist Episcopal Church, 1932). Among the Holiness theologians to cite Clarke was Wiley, *Christian Theology*, 1:459–461. Andrew Johnson, in his series "Methodism and Modern Thought," *Pentecostal Herald* 33 (July 20, 1921): 4, cited Clarke to support the Mosaic authorship of the Pentateuch. Randy L. Maddox kindly directed me to several pertinent sources.

53. Glenn Petersen, "How Did the Mammoths Get to Alaska?" *Light and Life* (February 1964): 8; Ira Edwards, "Science Won't Prove Faith," ibid. (February 1964): 14–15. See also the cluster of articles on creationism in *Good News* 15 (March/April 1982), published by evangelical Methodists in Wilmore, Kentucky: A. B. Broderson, "What's All This Fuss about Evolution and Creation?" pp. 8–17; Cecil B. Hamann and J. Paul Ray, "Progressive Creationism," pp. 12–14; and James V. Heidinger II, "Church and State: How Much Separation?" pp. 31–35.

54. Finis Jennings Dake, *Dake's Annotated Reference Bible* (Lawrenceville, GA.: Dake Bible Sales, 1963), pp. 54–57; Jimmy Swaggart Ministries, *The Pre-Adamic Creation and Evolution* (Baton Rouge, LA: Jimmy Swaggart Ministries, 1986); Lester Sumrall, *Genesis: Crucible of the Universe* (South Bend, IN: LESEA Publishing, 1982), p. 23; Kenneth E. Hagin, *The Origin and Operations of Demons*, 2nd ed. (Tulsa, OK:

Kenneth Hagin Ministries, 1987), p. 5; Gordon Lindsay, *The Bible Is a Scientific Book* (Dallas: Christ for the Nations, 1971), pp. 20–28; Gordon Lindsay, *Evolution—The Incredible Hoax* (Dallas: Christ for the Nations, 1973; originally printed in 1961); Charles F. Parham, *A Voice Crying in the Wilderness,* 4th ed. (Baxter Springs, KS: Robert L. Parham, 1944), pp. 81–85. On Dake, see P. H. Alexander, "Finis Jennings Dake (1902–87)," in *Dictionary of Pentecostal and Charismatic Movements,* pp. 235–236. On the continuing popularity of the gap theory, see Tom McIver, "Formless and Void: Gap Theory Creationism," *Creation/Evolution* 24 (Fall 1988): 1–24. Henry M. Morris's view of charismatic behavior is documented in his letter to J. C. Whitcomb, Jr., October 3, 1964, courtesy of John C. Whitcomb, Jr.

55. John C. Whitcomb, Jr., "How Did God Make Man?" *Pentecostal Evangel,* April 23, 1967, pp. 12–13, 27; "Researchers Show Earth Is Not Billions of Years Old," ibid., October 29, 1978, p. 24; H. Wayne Hornbaker, "Science Supports the Bible," ibid., September 14, 1980, pp. 8–9; "Former Evolutionist Says Many Scientists Are Helping Spur 'Modern-Day' Revival," ibid., May 2, 1982, p. 12; Paul L. Walker, *Understanding the Bible and Science* (Cleveland, TN: Pathway Press, 1976), pp. 44–49; Wade H. Phillips, *God, the Church, and Revelation* (Cleveland, TN: White Wing Publishing House, 1986); Dennis Lindsay, unpublished syllabus for the course "Scientific Creationism," Christ for the Nations Institute, 1986; Winkie Pratney, *Creation or Evolution?* (Glendale, CA: Church Press, n.d.); Winkie Pratney, "Creation or Evolution?" in *The Last Days Collection: A Treasury of Articles from Last Days Ministries* (Lindale, TX: Pretty Good Publishing, 1988), pp. 162–187. Pratney's scientific training is mentioned on p. 14 of *Last Days Collection.* I am indebted to Paul Boyer for this edition of Pratney's work, which also circulated as a series of tracts. The Assemblies of God position on creation appears in a statement adopted by the General Presbytery in 1977 and published in pamphlet form as *The Doctrine of Creation* (Springfield, MO: Gospel Publishing House, 1977).

56. Robert L. Selle, "No Evolute," *Pentecostal Herald* 38 (September 22, 1926): 6; H. C. Morrison, quoted in Wesche, *Henry Clay Morrison,* p. 143; Johnson, "The Evolution Articles," p. 6.

ACKNOWLEDGMENTS

During my graduate school days at the University of California at Berkeley my mentor, A. Hunter Dupree, introduced me to the world of Darwinian scholarship, and his influence has remained with me to the present. Over the years I have also benefited greatly from the writings and advice of my Darwinian friends David Livingstone, James Moore, John Stenhouse, and, especially, Jon Roberts, to whom I owe more than I care to acknowledge and to whom I dedicate this book as partial payment. David Hollinger and Charles Rosenberg (as well as Roberts) read the entire manuscript, and numerous other friends and colleagues read individual chapters (where they are acknowledged by name). David Lindberg encouraged me; Lester Stephens and Frank Sulloway collaborated with me; Michael Spierer advised me; Paul Boyer offered me sanctuary at the University of Wisconsin Institute for Research in the Humanities in the spring of 1997; Jack Ellis procured a copy of the message on evolution from the Alabama State Board of Education for me; Alex Holzman, of Cambridge University Press, reluctantly granted me an editorial dispensation; Louise E. Robbins read page proofs; Richard E. Davidson prepared the index. My sister, Carolyn Remmers, my niece, Sue Welke, and my friends Vern and Barbara Carner, Jon and Sharon Roberts, and Rennie and LeAnn Schoepflin cared for me. My favorite Darwinist, Karen Steudel, loved me. My daughter, Lesley Anne Numbers, gave meaning to my life.

RLN
New Year's Eve, 1997

INDEX